Sharpening Your Advanced SAS® Skills

Sharpening Your Advanced SAS® Skills

Sunil Gupta

CRC Press is an imprint of the
Taylor & Francis Group, an **informa** business
A CHAPMAN & HALL BOOK

CRC Press
Taylor & Francis Group
6000 Broken Sound Parkway NW, Suite 300
Boca Raton, FL 33487-2742

Printed on acid-free paper
Version Date: 20150831

International Standard Book Number-13: 978-1-4822-4037-5 (Hardback)

Visit the Taylor & Francis Web site at
http://www.taylorandfrancis.com

and the CRC Press Web site at
http://www.crcpress.com

Contents

List of tables

Preface

This book is a result of more than 20 years of SAS programming in the pharmaceutical industry. Based on my insights and experience, I created a reference of SAS programming techniques to become more organized and productive. Advanced SAS procedures and tools, such as Proc SQL and SAS Macro programming, need to be mastered in any industry. Improving program efficiency and taking advantage of the latest SAS options and new SAS procedures also play an important role in advancing your career.

In general, chapters are organized to facilitate easy searching of key points by summarizing and differentiating the syntax between similar SAS statements and options. Once found, the real-world SAS examples can be copied to your working program. The basic syntax for Proc SQL, for example, will be explained and illustrated with simple, common task-oriented examples. Mindmaps on SASSavvy.com and process flowcharts help to communicate concepts and relationships. Questions will be included at the end of each section to reinforce your knowledge of the topic.

At the beginning of each chapter, a chapter overview is provided to facilitate quick reference to the detailed examples and syntax in the chapter. The basic syntax, expected data, and descriptions are organized in summary tables to facilitate memory recall of the information. General rules within each section list common points about similar statements or options. Each topic includes the basic syntax, a series of key points, and other notes about the specific SAS statement or options.

Examples of SAS program and code statements are line numbered with references, such as SAS papers and websites, for more detailed explanations. Note that SAS examples are a complete block of code that can be executed, while selected SAS code syntax needs to be executed as part of the remaining program. This unique approach empowers both the advanced programmer who needs a quick refresher as well as programmers interested in learning new programming techniques. By studying and answering the section questions, you will be better prepared for the advanced SAS certification exam.

Acknowledgments

As I come to the end of writing the book *Sharpening Your Advanced SAS® Skills*, I would like to thank my wife Bindiya, and daughters, Aarti and Anupama. Their excitement, encouragement and support are greatly appreciated. I also want to thank the many university student members of SASSavvy.com for their quest for greater SAS insights which helped add to the over 200 common frequently asked SAS technical questions database.

I would also like to thank SAS technical support for its valuable feedback, especially on the more advanced topics such as hash table objects. I hope this book's organization and content will inspire you to reach higher SAS technical goals such as better preparation for the advanced SAS certification exam or becoming a stronger contributor to your organization's SAS development environment. Happy SAS programming.

About the author

Sunil Gupta is an international speaker, best-selling SAS author, and global corporate trainer. He is a principal SAS/CDISC consultant. Most recently, he taught both of his CDISC online classes with the University of California at San Diego and SAS Institute India. In 2011, Gupta launched his unique SAS resource blog, SASSavvy.com, for smarter SAS searches. Currently, SASSavvy.com's membership consists mostly of SAS programmers, university students, and pharmaceutical corporate accounts.

Because of Sunil's mentoring and contributions to the pharmaceutical industry as well as his commitment to high quality standards, Sunil was recently asked to join The Science Advisory Board. In 2013, Gupta was recognized by SAS Institute's Circle of Excellence for 20 years of service. In 2008, he was chosen as one of the "100 Notable People in the Medical Device Industry" for his contributions. Two of his popular pharmaceutical industry leadership articles include "How Cloud-Based Tools Can Help with FDA Compliance" in *Life Science Leadership* magazine and "Standards for Clinical Data Quality and Compliance Checks" in the journals *Society for Clinical Data Management* and *Pharmaceutical Programming.*

Each year, Gupta has been an invited presenter at many SAS conferences for his "highly acclaimed" Proc SQL hands-on workshop. Sunil has also contributed three popular SAS blogs—When do I use a WHERE statement instead of an IF statement to subset a dataset?; Something for nothing? Adding group descriptive statistics; and Converting variable types—use PUT() or INPUT()? He has been using SAS® software for over 20 years and is a SAS Base Certified Professional. He is also the author of *Quick Results with the Output Delivery System* and *Sharpening Your SAS® Skills.*

Accessing Data Using SQL

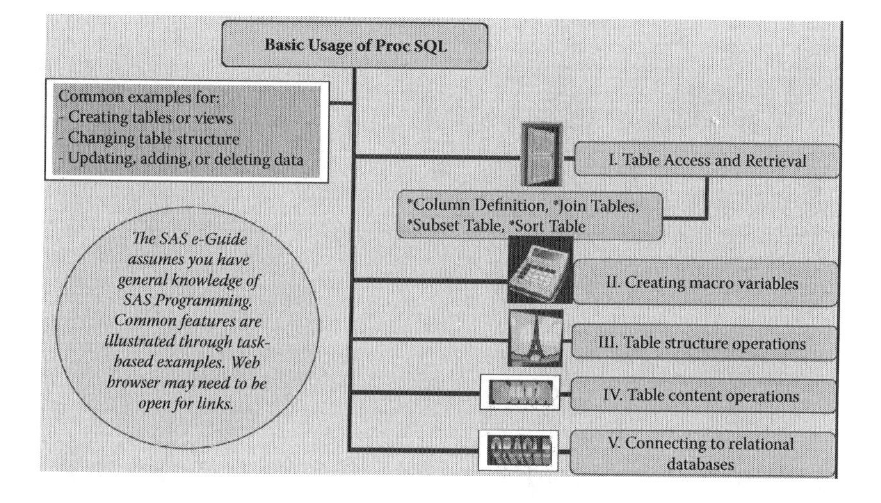

Chapter Overview

1.1 INTRODUCTION

This chapter is organized to facilitate easy searching of key points by summarizing and differentiating the syntax between similar SAS statements and options. The basic syntax for PROC SQL will be explained and illustrated with simple task-oriented examples. Questions will be included at the end of each section to reinforce the user's knowledge of the topic.

At the beginning of the chapter, a chapter overview is provided to facilitate quick reference to the detailed examples and syntax in the chapter. The basic syntax, expected data, and descriptions are organized in summary tables to facilitate memory recall of the information. General rules within each section list common points about similar statements or options. Each topic includes the basic syntax, a series of key points, and other notes about the specific SAS statement or options. Examples of SAS program and code statements are line numbered with references for more detailed explanation. Note that SAS examples are a complete block of code that can be executed, while selected SAS code syntax needs to be executed as part of the remaining program. This unique approach empowers both the advanced programmer who needs a quick refresher as well as programmers interested in learning new programming techniques.

1.1.1 Proc SQL Basic Usage

```
1. proc sql;                        /* required */
2.  < proc sql statement 1 > ;
3.  < proc sql statement 2 > ;
4. quit;                            /* required */
```

Lines 2, 3: All of the examples in the e-Guide are expected to be embedded between **lines 1** and **4**. Along with the required SAS statements within each PROC SQL block of code, SAS also accepts one or more of the following optional statements.

1.1.2 Eight Common Benefits for Using PROC SQL

```
1.  proc sql;                         /* required */
2.  create table mytable as select ...; /* 1 - new table */
3.  create view myview as select ...;  /* 2 - new view */
4.  select ...;                        /* 3 - powerful query tool */
5.  alter table mytable ...;           /* 4 - new table structure */
6.  update table mytable ...;          /* 5 - update table content */
7.  insert table mytable ...;          /* 6 - update table content */
8.  delete from table mytable ...;     /* 7 - delete table content */
9.  drop table mytable;                /* 8 - delete table */
10. quit;                              /* required */
```

Line 1: Six options exist to help debug — NOEXEC (to check syntax without executing the code), FEEDBACK (displays code executed), INOBS = # of rows read, OUTOBS = # of rows written, _METHOD (displays PROC SQL execution options), _TREE (display visual structure of logic). See SAS Online reference for a complete set of features and options [1]. See SAS paper on undocumented features [2]. Remember to set OPTIONS MSGLEVEL = I for _METHOD and _TREE options, since default is MSGLEVEL = N. After the syntax is error-free, you may need to investigate your data and logic. Note that once debug options are applied, the RESET NOFEEDBACK is required, for example, to reset to default. NUMBER option displays the row number, which can also be saved as a new column with the MONOTONIC() keyword. In DATA step, the MONOTONIC()

is similar to _N_. See SAS paper on the MONOTONIC() keyword [3]. The NOWARN option can be added to prevent displaying warning messages. Finally, the FLOW option is useful to wrap text within each cell.
Line 2: Create table MYTABLE with SELECT statement. If a table already exists, then LIKE, instead of AS, <table> option can be used to copy the table structure with zero rows.
Line 3: Create view MYVIEW with SELECT statement. Views are executable instructions in the form of a SELECT statement to extract data from one or more tables. Views are similar to tables except that views do not contain data so do not require space and cannot be indexed. Once created, it is useful to DESCRIBE view to display the table structure, similar to PROC CONTENTS, to the SAS log.
Line 4: Query table with SELECT statement. Add VALIDATE before SELECT to check syntax without executing the code. Note that while generally PROC SQL is used for data management tasks, the results of SELECT statements may be directed to Excel, RTF, PDF, or HTML files through ODS. See SAS site for overview of SELECT statement [4]. Also note that PROC SQL stores the number of rows returned in the &SQLOBS automatic macro variable that can be used for further macro processing. See automatic macrovariable section. See SAS site for more information on automatic macrovariables [5].
Line 5: Change MYTABLE table structure by copying another table, creating columns with attributes, dropping columns, modifying column attributes, or adding integrity constraints. See Section 1.4 Table Structure Operations.
Line 6: Update MYTABLE table content by multiplying by number or string modification. Changes can be conditionally applied with the WHERE clause. See Section 1.4 Table Content Operations.
Line 7: Add data content to MYTABLE table as series of values for all columns, one column or data from another table. See Section 1.4 Table Content Operations.
Line 8: Delete table content from MYTABLE table based matching condition from WHERE clause.
Line 9: Delete table MYTABLE. See Section 1.4 Table Content Operations.
Line 10: QUIT is required.

1.2 TABLE ACCESS AND RETRIEVAL

Creating SAS tables enables quick access to the data. Options for data access and retrieval include selecting columns, joining tables, subsetting tables, and sorting tables.

1.2.1 Basic Lines of Code for Table Access and Retrieval — Six Components

```
1. create table mytable as
2. select name, sex, max(height) as mhgt   /* select columns - Step 5 */
3. from sashelp.class                      /* source table - Step 1 */
4. where sex =        'F'                  /* option to subset table - Step 2 */
5. group by sex                            /* option to group by - Step 3 */
```

```
6. having weight > avg(weight)      /* option to include having - Step 4 */
7. order by name                    /* option to sort table - Step 6 */
8. ;                                /* required */
```

Line 1: Create table MYTABLE. As in the DATA Step, you can also specify dataset options such as DROP = or KEEP =. Note that all six PROC SQL components are contained within one statement and must be specified in this order: 1) selecting, 2) source table, 3) row subset condition, 4) group by, 5) group subset condition, 6) sort order. Four additional dimensions of PROC SQL over the DATA Step include: 1) descriptive statistics using summary functions along with COALESCE(), 2) group by columns, 3) conditional processing, and 4) row and group level subsetting along with subqueries.
Line 2: Select or create columns. See Column Definition.
Line 3: From source table. See Join Tables.
Line 4: Apply detail level condition for subsetting table. See Subset Tables.
Line 5: Can group by column.
Line 6: Can apply group level condition for subsetting based on summary function.
Line 7: Sort result table. See Sort Table.
Line 8: ';' required to end PROC SQL statement.

```
SAS Output
                  Obs   Name      Sex   mhgt
                   1    Barbara    F    66.5
                   2    Carol      F    66.5
                   3    Janet      F    66.5
                   4    Mary       F    66.5
```

1.2.2 Column Definition

When defining columns, you can select columns that already exist in the table or create new columns. When creating new columns, remember to specify column attributes, such as label, format, and length. Finally, when selecting or creating columns to be saved as a table, the column order defines the order stored in the table.

Note that as opposed to the DATA step, the new column name must be different from the existing column names. Note also that if any new column created within the same PROC SQL step is specified again, then the CALCULATED keyword must be placed before the new column name such as in the WHERE clause. Use caution when specifying the new column name again within the same select clause. See Subset Tables.

1.2.2.1 Four options for selecting columns: select clause

```
1. select name, sex
2. select name label = "My Label" format = $10. length = 10
3. select *
4. select distinct sex        /* alternative to use unique */
```

Line 1: List selected columns, separate each column by comma ',' and in order of output.
Line 2: After the column name, to define column attributes such as label, format, and length. Note that LENGTH option for character columns is still without '$' and

that CHAR is required in the CREATE TABLE statement. LENGTH is always equal to a number. See Table Structure Operations.
Line 3: Use wildcard '*' to select all columns from the table.
Line 4: Assure unique values for SEX column. Note that DISTINCT applies to all combinations of columns in the SELECT statement. As an alternative, UNIQUE can be applied. Note that missing values are also considered unique and will appear.

1.2.2.2 Ten options for creating columns: select clause

```
1. select int((age+150)/10) as myage          /* AS keyword is required */

2. select max(height, weight) as maxval label = 'Max Val' length = 4

3. select max(max(height, weight)) as grpmaxval label = 'Max Val' length = 4

4. select sex, weight, ((weight/sum(weight))*100) as wpercnt length = 4
     from sashelp.class group by sex;

5. select 'my constant' as myname label = 'My Name' length = 11 format = $11.

6. select trim(left(city))||","||trim(left(state)) as address format = $45.

7. select case      /* any valid expression after when and then */
     when age > 0 and age < 13    then 1
     when age between 13 and 15 then 2 /* 13 & 15 are inclusive */
     when age > 15        then 3
     else             .         /* best practice */
   end as agegrp           /* create numeric or character column */

8. select case sex           /* SEX specified only once after CASE */
     when 'M' then 1      /* similar to when sex = 'M' */
     when 'F' then 2      /* similar to when sex = 'F' */
     else      .
   end as sexnum

9. select sum(sex = 'M') as nmale, sum(sex = 'F') as nfemale

10. create table class2 as
     select a.*, b.sexn from sashelp.class as a
     left join
     (select name, count(sex) as sexn from sashelp.class where sex > ' '
     group by sex)
     as b on a.name = b.name;
```

Line 1: Use the INT() function in a calculation expression. AS keyword followed by the new column name is required. SAS functions can be nested. Almost any SAS function [6] can be specified. Arithmetic operations include +, -, *,/, **. Note that with the AS option, you can rename an existing column.
Line 2: Use summary MAX() function to summarize across columns HEIGHT and WEIGHT, which is an example of across-the-record processing. Other summary functions include AVG(), COUNT(), MIN(), NMISS(), STD(), and VAR(). The COUNT() function excludes missing values, while the other summary functions include missing values unless a WHERE clause is applied. Use caution when applying the UNIQUE or

DISTINCT option. Set UNIQUE just before the variable and summary function as the outer function, such as COUNT(UNIQUE DOSEDT). See SAS website for full list of summary functions and summary function tips [7]. See section on useful PROC SQL functions.

Line 3: Nested summary functions allow for both across-the-columns and then down-the-column calculations. First, the maximum value of height and weight is saved, and then the maximum value across all rows is saved as a single value.

Line 4: Use summary statistics SUM() function to summarize the WEIGHT column by SEX. This is an example of down-the-record processing. This technique is generally used with GROUP BY to calculate summary statistics by group instead of by all rows. Note that it is possible to calculate group descriptive statistics on variables different from the GROUP BY variable. With the WEIGHT column also selected, this is a useful option for adding summary statistics back to the original data — called "re-merging." Generally, the granularity of the results is determined from the columns selected. To repeat this in a DATA step may require multiple DATA steps.

Line 5: Use a character constant column equal to 'my constant.' It is best to include LENGTH and FORMAT options. Option includes specifying a numeric constant, such as 150 or '01JAN2010'd.

Line 6: Use TRIM() and LEFT() character functions in a character expression with CITY and STATE columns to concatenate without blanks. It is best to include LENGTH and FORMAT options.

Line 7: Use mutually exclusive WHEN and THEN conditions based on any valid expression with ELSE to set to 0. Each row is evaluated as true, false, or unknown. Default is '.' for numeric columns. Can use LENGTH option after AGEGRP to override default length of 8. END AS is required for CASE clause. Also applicable for creating new character columns, default length of 25. Note the summary functions may not work as expected in conditional THEN clause. Instead, apply summary functions on the complete CASE clause. Note also that THEN-DO-END blocks do not work as they do with DATA Step programming.

Line 8: This is an alternative to method 6 that is useful to save typing time. It can apply only for simple value comparisons such as 'M', 'F', or 1, 2. Note that this is the only method for referencing CALCULATED new columns in the CASE clause. **Line 7** will not work for CALCULATED new columns.

Line 9: NMALE stores the number of males and NFEMALE stores the number of females. The 'SEX = 'M'' is another syntax for where conditions. Apply with caution since error may be generated.

Line 10: A more effective technique for adding group descriptive statistics back to the original dataset is to apply subquery and summary functions by the group variable. SEX count values are added back to the dataset without rearranging the sort order of the dataset. This technique can be applied to also save MIN, MAX, etc., of the summary function variable. See SAS tip for more information [8].

1.2.2.3 SAS output for line 4

```
Sex     Weight     wpercnt
M           83    7.618173
M        102.5    9.407985
M         99.5     9.13263
M          112    10.27994
M        112.5    10.32584
```

```
M           84    7.709959
M           85    7.801744
M          133    12.20743
M          128    11.74851
```

1.2.2.4 SAS output for line 10

Obs	Name	Sex	Age	Height	Weight	sexn
11	Joyce	F	11	51.3	50.5	9
12	Judy	F	14	64.3	90.0	9
13	Louise	F	12	56.3	77.0	9
14	Mary	F	15	66.5	112.0	9
15	Philip	M	16	72.0	150.0	10
16	Robert	M	12	64.8	128.0	10
17	Ronald	M	15	67.0	133.0	10
18	Thomas	M	11	57.5	85.0	10
19	William	M	15	66.5	112.0	10

1.2.3 Join Tables

Joining tables is easy to accomplish with PROC SQL. Prior to SAS V9, you could join up to 32 tables. After SAS V9, however, you can join up to 256 tables with options for inner and outer joins. Inner joins, or equijoins, return a table containing rows that match both tables. Outer joins return a table containing rows that match both tables plus all nonmatching rows from the LEFT, the RIGHT, or both tables. Use caution when applying outer joins on blocks of two tables at a time. Note that dataset options, such as RENAME, KEEP, or DROP, are still valid with PROC SQL. See graphic SAS paper on PROC SQL Joins [9].

Note that in the SELECT clause, one can use COALESCE(CLASS.NAME, STUDENTS.NAME) function with FULL JOIN to resemble the MERGE statement with storage of nonmissing NAME value from either CLASS as first selection or STUDENTS table as second selection. Note that default of PROC SQL is to store left dataset values for common nonby variables. For single tables, COALESCE(CLASS. NAME, 'unknown') is also useful to prevent missing values. Finally, make sure to specify table names before accessing variables when joining my multiple key variables.

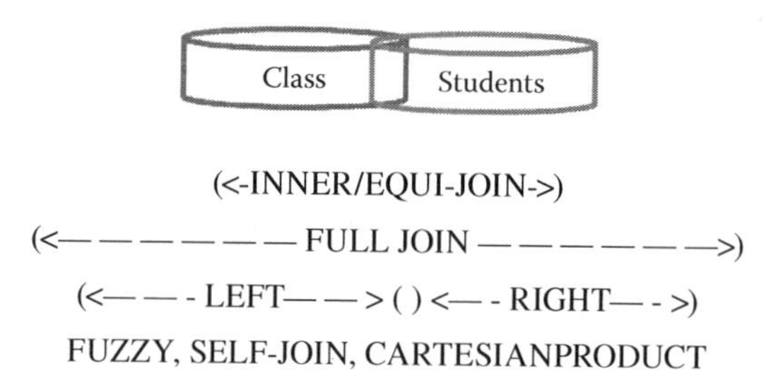

1.2.4 Seven Options for Horizontal Joins (Different Table Structures)

1.2.4.1 Basic lines of code for inner/equi-join (matching rows)

```
1. create table mytable as
2. select class.name, students.sex
3. from sashelp.class as class
4. ,                 /* code for joining 2 tables */
5. mylib.students as students
6. where class.name = students.name and class.sex = students.sex
7. ;      /* option to add subset conditions or order by clauses */
8. create table mytable as
9. select class.name, students.sex, students2.sex as sex2
10. from sashelp.class as class
11. ,                /* code for joining more than 2 tables */
12. mylib.students as students
13., mylib.students as students2
14. where class.name = students.name and class.sex = students.sex
15. and class.sex = students2.sex
16. ; /* lines 13 and 15 are repeated for each table */
```

Alternative syntax to WHERE clause:

```
17. create table mytable as
18. select class.name, students.sex
19. from sashelp.class as class
20. inner join            /* code for joining 2 tables */
21. mylib.students as students
22. on class.name = students.name and class.sex = students.sex
23. ; /* optional WHERE subset condition clause can also be applied */

24. create table mytable as
25. select class.name, students.sex, students2.sex as sex2
26. from sashelp.class as class
27. inner join            /* code for joining more than 2 tables */
28. mylib.students as students
29. on class.name = students.name and class.sex = students.sex
30. inner join            /* lines 30, 31 and 32 are repeated for each table */
31. mylib.students as students2
32. on class.sex = students2.sex
33. ;
```

Line 1: Create MYTABLE defined by the select results.
Line 2: Select NAME from CLASS and SEX from STUDENTS tables. Any column from any table in the FROM clause can be selected. CLASS.*, STUDENTS.* will select all columns from both CLASS and STUDENTS tables. This results in overwrite messages, since the NAME and SEX columns are common. See section on COALESCE() to help resolve this. Note that the table name prefix is only required when tables have common column names.

Line 3: Use SASHELP.CLASS as alias name CLASS; alias can be used with columns to reference table source. To take advantage of presorted tables, then the (SORTEDBY = <key column>) option can be applied.
Line 4: Comma ',' is required to separate tables. Default is inner join without any operator.
Line 5: Use MYLIB.STUDENTS table as alias name STUDENTS.
Line 6: Conventional inner join — only the matching rows are selected. To merge by matching NAME and SEX columns between CLASS and STUDENTS tables. Note that NAME and SEX could have different names. Note that without a WHERE clause, a Cartesian product is created, which causes SAS to link all rows from one table with all rows from another table or a many-to-many join. Finally, fuzzy or sounds-like matches (similar to many-to-many join) are also possible for close, but not necessarily exactly equal matches. Fuzzy joins can be performed on range of values using BETWEEN, closeness of character columns using SOUNDEX(), or joins with additional conditions.
Line 7: Along with the linking WHERE clause, additional subset conditions can also be applied. Note that all of these clauses are in one PROC SQL statement.
Line 8–Line 16: Code for joining more than 2 tables. **Lines 13** and **15** are repeated for each new table joined.
Line 17: Create mytable defined by the select results.
Line 18: Select NAME from CLASS and SEX from STUDENTS tables.
Line 19: Use SASHELP.CLASS as alias name CLASS; alias can be used with columns to reference table source.
Line 20: INNER JOIN is similar to WHERE clause.
Line 21: Use MYLIB.STUDENTS table as alias name STUDENTS.
Line 22: ON is required for inner joins to merge by matching NAME and SEX columns between CLASS and STUDENTS tables. Note that SAS functions are also acceptable on ON clauses.
Line 23: Optional WHERE can also be applied to subset table.
Line 24–Line 33: Code for joining more than two tables. **Lines 30, 31,** and **32** are repeated for each new table joined.

1.2.4.2 SAS output for lines 1–7

```
Obs    Name       Sex
 11    Joyce       F
 12    Judy        F
 13    Louise      F
 14    Mary        F
 15    Philip      M
 16    Robert      M
 17    Ronald      M
 18    Thomas      M
 19    William     M
```

1.2.4.3 Basic lines of code for outer join (LEFT, FULL, or RIGHT)

1. create table MYTABLE as
2. select class.name, students.sex

```
3. from sashelp.class as class        /* code for joining 2 tables */
4. left join                          /* left, full, or right join */
5. mylib.students as students
6. on class.name = students.name and class.sex = students.sex
7. ;      /* optional WHERE subset condition clause can also be applied */

8. create table mytable as
9. select coalesce(class.name, students.name) as name,
10. students.sex, students2.sex as sex2
11. from sashelp.class as class
12. left join          /* code for joining more than 2 tables */
13. mylib.students as students
14. on class.name = students.name and class.sex = students.sex
15. left join          /* lines 15, 16 and 17 are repeated for each table */
16. mylib.students as students2
17. on class.sex = students2.sex
18. ;
```

Line 1: Create MYTABLE defined by the select results.
Line 2: Select NAME from CLASS and SEX from STUDENTS tables.
Line 3: Use SASHELP.CLASS as alias name CLASS; alias can be used with columns to reference table source. Note that ',' is excluded in outer joins.
Line 4: From outer LEFT JOIN to keep all CLASS rows and only matching STUDENTS rows, which may cause missing values in the STUDENTS table columns. Options include RIGHT JOIN or FULL JOIN to keep all STUDENTS rows and only matching CLASS rows or rows from both CLASS and STUDENTS tables correspondingly.
Line 5: Use MYLIB.STUDENTS table as alias name STUDENTS.
Line 6: ON is required for outer joins to merge by matching NAME and SEX columns between CLASS and STUDENTS tables.
Line 7: Optional WHERE can also be applied to subset table.
Line 8–Line 18: Code for joining more than two tables. **Line 9** uses COALESCE() function to select the first nonmissing NAME. **Lines 15, 16,** and **17** are repeated for each new table joined.

Pulling it all together: Example of summary to detail issues listing from multiple datasets

```
* create population dataset as non-missing flags from each dataset based on condition,
  summarize by pt;
* create selected vars dataset with merge back to display details, add more vars as
  needed;

   proc sql;
    create table rdatapop as
    select unique coalesce(a._pt_, b._pt_, c._pt_, d._pt_) as pt,
     ifn(a.randdt > ., 1, .) as franddt,
     max(ifn(b.dosetl > ., 1, .)) as fdosetl,
     ifn(c.EOT > '', 1, .) as feot,
     ifn(d.eos > '', 1, .) as feos
    from sdfraw.enrlment_all as a
```

```
  full join sdfraw.ipadmin_all as b on a._pt_=b._pt_
  full join sdfraw.eot_all as c on a._pt_=c._pt_
  full join sdfraw.eos_all as d on a._pt_=d._pt_
  group by calculated pt;

  create table rdata as
  select unique coalesce(a._pt_, b._pt_, c._pt_, d._pt_) as pt, z1.*,
   a.randdt,
   b.dosetl, b.STARTDT, b.dosetl, b.dose_u_std, b.folder,
   c.EOT,
   d.eos
  from rdatapop as z1
  left join sdfraw.enrlment_all as a on z1.pt=a._pt_
  left join sdfraw.ipadmin_all as b on a._pt_=b._pt_
  left join sdfraw.eot_all as c on a._pt_=c._pt_
  left join sdfraw.eos_all as d on a._pt_=d._pt_
  group by calculated pt;
quit;

* Summary review;
* sequence order;
* 1. franddt - randomized;
* 2. fdosetl - received dose;
* 3. feot - end of treatment reason;
* 4. feos - end of study reason;
proc freq data=rdatapop;
 tables franddt*fdosetl*feot*feos/list missing;
run;

* Detail review - display all vars based on issue condition;
proc print data=rdata noobs;
 where franddt=1 and fdosetl=.;
 sum franddt fdosetl feot feos;
run;
```

1.2.4.4 Basic lines of code for FUZZY-join

```
1. create table mytable as
2. select class.name, students.sex
3. from sashelp.class as class
4. ,
5. mylib.students as students
6. where soundex(class.name) = soundex(students.name);
```

Line 1: Create table MYTABLE defined by the select results.
Line 2: Select NAME from CLASS and SEX from STUDENTS tables.
Line 3: Use SASHELP.CLASS as alias name CLASS; alias can be used with columns to reference table source.
Line 4: Comma ',' is required to separate tables.
Line 5: Use MYLIB.STUDENTS table as alias name STUDENTS.

Line 6: While SOUNDEX() is only required on one side of ' = ', having SOUNDEX() on both sides of ' = ' will provide greater flexibility. Fuzzy joins enable use of BETWEEN operator for range joins, such as WHERE A.WEEKBEG BETWEEN B.BEGWEEK AND B.ENDWEEK, SPEDIS() and SOUNDEX() function for closeness of character values and HAVING clause for joining by exact matches and detail or summary level function. The syntax for applying SPEDIS() is SPEDIS(CLASS.NAME, STUDENTS.NAME) < 20, for example, to expand list of possible matches. In general, values greater than 20 pick up mostly nonmatches. You may also consider the colon modifier: EQT, GTT, LTT, GET, LET, and NET. Note that this example uses WHERE clause; fuzzy joins can also be performed with outer joins using the ON clause.

1.2.4.5 Basic lines of code for SELF-join

```
1. create table mytable as              /* useful for internal comparison */
2. select a.name, a.sex as sex_a, b.sex as sex_b
3. from sashelp.class as a
4. ,                               /* similar to many-to-many join */
5. sashelp.class as b
6. where a.age < = b.age;         /* WHERE can compare same column name */
```

Line 1: Create table MYTABLE defined by the select results.
Line 2: Select NAME from A and SEX from A and B.
Line 3: Use SASHELP.CLASS as alias name A; alias can be used with columns to reference table source.
Line 4: Comma ',' is required to separate tables.
Line 5: Copy of SASHELP.CLASS is made with B alias.
Line 6: WHERE clause allows complex condition of same column from the two tables, which is generally not possible with the basic PROC SQL joins.

1.2.4.6 SAS output

```
Obs   Name      sex_a   sex_b
211   Louise      F       M
212   Mary        F       M
213   Robert      M       M
214   Ronald      M       M
215   Thomas      M       M
216   William     M       M
```

1.2.4.7 Basic lines of code for CARTESIAN PRODUCT

```
1. create table mytable as
2. select class.name, students.sex
3. from sashelp.class as class
4. ,                              /* many-to-many join */
5. mylib.students as students;       /* no WHERE clause */
```

Line 1: Create table MYTABLE defined by the select results.
Line 2: Select NAME from CLASS and SEX from STUDENTS tables.

Line 3: Use SASHELP.CLASS as alias name A; alias can be used with columns to reference table source.
Line 4: Comma ',' is required to separate tables.
Line 5: Without a WHERE clause, a Cartesian product, many-to-many join, is performed. Use caution when joining without a WHERE clause. This technique is often used to join records by date ranges, such as adverse and conmeds and is similar to fuzzy joins. Note that for all joins, SAS first creates a Cartesian product before subsetting the table with the WHERE clause.

As an alternative to joining tables horizontally, you can also join tables vertically using set operators: UNION (unique LEFT and RIGHT), OUTER UNION CORR (all LEFT and RIGHT), INTERSECT (matching LEFT and RIGHT), and EXCEPT (LEFT not in RIGHT of common columns). By default, UNION, INTERSECT, and EXCEPT return unique rows.

1.2.5 Four Options for Vertical Joins/Set Operators

Classes A and B have similar Table Structures.

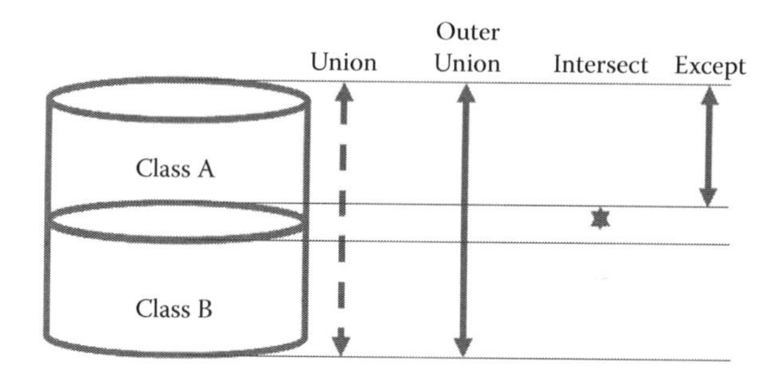

(UNION – Unique, A and B)

(OUTER UNION CORR – All A and B)

(INTERSECT – Unique Matching A and B)

(EXCEPT – Unique A not in B of common columns)

Four Options for Vertical Joins (Similar Table Structures)
Limited to joining two tables at a time

1.2.5.1 *Basic lines of code for UNION join*

```
1. create table mytable as
2. select name, sex from sashelp.class          /* select similar columns */
3. union                         /* selects all unique rows */
4. select name, sex from mylib.students;
```

Line 1: Create table MYTABLE defined by the select results.
Line 2: Select NAME and SEX columns from SASHELP.CLASS. It is best to select only common columns between SASHELP.CLASS and MYLIB.STUDENTS

instead of '*' for all columns to assure correct number of unique records are selected. Note that this SELECT clause can be used as a subquery to select more variables once the key variables are identified.
Line 3: Nonmatching rows from both SASHELP.CLASS and MYLAB.STUDENTS are selected. The ALL keyword is an option to include all rows instead of just unique rows.
Line 4: Select NAME and SEX columns from MYLIB.STUDENTS. It is best to select common columns selected from CLASSA.

1.2.5.2 Basic lines of code for OUTER UNION CORR join

```
1. create table mytable as
2. select name, sex from classa    /* select similar columns */
3. outer union corr                /* selects all rows from both tables */
4. select name, sex from classb;
```

Line 1: Create table MYTABLE defined by the select results.
Line 2: Select NAME and SEX columns from CLASSA. It is best to select only common columns between CLASSA and CLASSB instead of '*' for all columns to assure correct number of unique records are selected. Note that this SELECT clause can be used as a subquery to select more variables once the key variables are identified.
Line 3: All rows from CLASSA and CLASSB are selected. Without the CORR keyword, similar name columns will not overlay each other and be displayed as separate columns. OUTER UNION CORR is similar to FULL JOIN and SET CLASSA CLASSB in DATA Step.
Line 4: Select NAME and SEX columns from CLASSB. It is best to select common columns selected from CLASSA.

1.2.5.3 Basic lines of code for INTERSECT join

```
1. create table mytable as
2. select name, sex from classa    /* select similar columns */
3. intersect    /* selects only unique matching rows from A and B tables*/
4. select name, sex from classb;
```

Line 1: Create table MYTABLE defined by the select results.
Line 2: Select NAME and SEX columns from CLASSA. It is best to select only common columns between CLASSA and CLASSB instead of '*' for all columns to assure correct number of unique records are selected. Note that this SELECT clause can be used as a subquery to select more variables once the key variables are identified.
Line 3: Unique matching rows from CLASSA and CLASSB are selected. The ALL keyword is an option to include all rows instead of just unique rows.
Line 4: Select NAME and SEX columns from CLASSB. It is best to select common columns selected from CLASSA.

1.2.5.4 Basic lines of code for EXCEPT join

```
1. create table mytable as
2. select name, sex from classa    /* select similar columns */
3. except    /* selects only unique non-matching rows from A table */
4. select name, sex from classb;
```

Line 1: Create table MYTABLE defined by the select results.
Line 2: Select NAME and SEX columns from CLASSA. Best to select only common columns between CLASSA and CLASSB instead of '*' for all columns to assure correct number of unique records are selected. Note that this SELECT clause can be used as a subquery to select more variables once the key variable values are identified.
Line 3: Unique nonmatching rows from CLASSA only are selected. The ALL keyword is an option to include all rows instead of just unique rows. This is useful to identify new or modified NAME or SEX values in CLASSA.
Line 4: Select NAME and SEX columns from CLASSB. It is best to select the common columns selected from CLASSA. Note that the UNIQUE option may be needed to return only one record per NAME and that the WHERE condition can be applied to restrict records.

1.2.6 Subset Tables

When subsetting tables, you can subset by rows or by groups. The difference is in the selection condition. Subsetting by row compares rows to an expression, and subsetting by group compares rows to a group expression based on summary functions. Apply caution when subsetting by both rows and by groups, since row subsetting occurs before group subsetting. Options for comparison operators include: =, ^ =, <, < =, >, and > =. Options for predicates include AND, BETWEEN, IN, IS NULL, IS MISSING, LIKE, and EXISTS. As an alternative to column names, references can be made to SELECT column order number, such as 1 and 2.

Subsetting the table using subqueries offers greater power and flexibility in preprocessing a table to dynamically specify the subset condition from another table or the same table. Subqueries offer a great alternative to break up a complex query if it is not possible to execute as one step. Examples show subqueries creating intermediate tables based on one column creating one or more values, as well as multiple columns creating multiple values. Note that keywords ANY and ALL can be specified for comparisons against a list of values after another operator is applied, such as (< or >). Correlated subqueries with EXISTS are subqueries referencing the same table as the outer table. Note that nested subqueries are not displayed.

1.2.6.1 Four options for subsetting by row

```
1. where weight = 150
2. where calculated myage < 50
3. where index(prodname, 'desk') > 0
4. where name eqt 'Jim'          /* comparable to = : in DATA Step */
```

Line 1: Condition based on WEIGHT expression.
Line 2: Condition based on MYAGE CALCULATED column. CALCULATED keyword is required before MYAGE if MYAGE column was created in the same PROC SQL step.
Line 3: Condition based on INDEX() function. Almost any SAS function can be applied.

Line 4: Almost any valid WHERE clause, including special operators, such as LIKE or BETWEEN, is acceptable. Note that colon (:) modifier counterpart also is acceptable as one of these syntax versions: EQT, GTT, LTT, GET, LET, and NET. See SAS paper for more info [10].

1.2.6.2 Option for subsetting by group

```
1. group by weight
2. having avg(weight) > 200
```

Line 1: Often used with summary functions in SELECT clause to summarize by columns. Required with HAVING clause. See Column Definition.
Line 2: Subset table based on grouping condition — i.e., not available in each row; requires a summary function and the GROUP BY clause. Use caution when applying summary functions with DISTINCT option. It can be useful, for example, to identify duplicate rows — SELECT KEY COLUMNS and GROUP BY NAME HAVING COUNT(DISTINCT NAME) > 1.

1.2.6.3 Option for subsetting by (1) row and then by (2) group

```
1. select sex, name, age, weight, avg(weight) as AVGWGT label='AVGWGT'
2. from sashelp.class where age >= 15 group by sex;
```

Lines 1–2: shows records after applying the WHERE condition to subset by detail records and group by sex.

```
1. select sex, name, age, weight               /* 4th step */
2. from sashelp.class
3. where age >= 15                  /* 1st step subsets by rows */
4. group by sex                  /* 2nd step groups rows by sex */
5. having weight > avg(weight);     /* 3rd step subset group averages */
```

Note that SAS processes the SELECT clauses in this sequence:

WHERE, GROUP BY, HAVING and then SELECT.

Line 1: SAS's fourth step is to select NAME, WEIGHT, and create AVG_WGT columns from SASHELP.CLASS.
Line 2: from SASHELP.CLASS table.
Line 3: SAS's first step is to subset rows for SEX = 'F' before applying any summary functions.
Line 4: SAS's second step is to GROUP BY SEX.
Line 5: SAS's third step is to subset groups by SEX and select only weights greater than group average. Note that if both WHERE (before) and HAVING (after) clauses are specified with the GROUP BY clause, then the row level subset condition is applied first, then any summary functions and then the group level subset condition is applied last. Note that it is acceptable to specify WHERE and GROUP BY clauses without including the HAVING clause.

1.2.6.4 SAS output

Sex	Name	Age	Weight
F	Janet	15	112.5
M	Philip	16	150
M	Ronald	15	133

1.2.6.5 Option for group summary function (1), subsetting by (2) group, and then subquery (3)

1. create table postm2 as
2. select a.*, b.aval, b.dtype, b.anl01fl
3. from postm (drop = aval) as a
4. right join
5. (
6. select unique usubjid, paramn, avisitn, mean(aval) as aval, 'AVERAGE' as dtype, 'Y' as anl01fl
7. from postm as a
8. group by usubjid, paramn, avisitn
9. having min(awtdiff) = awtdiff
10.)
11. as b on a.usubjid = b.usubjid and a.paramn = b.paramn and a.avisitn = b.avisitn
12. Having; alone in the last line makes it easier to see the end of the SELECT statement.

Line 2: Select all a.* and new b variables.
Line 3: Drop variable being averaged.
Line 4: RIGHT JOIN to keep only the matching record so that POSTM2 dataset can be appended to POSTM dataset to create new 'AVERAGE' records.
Lines 5–10: Subquery
Line 6: Select unique matching variables, USUBJID, PARAMN, and AVISITN, along with MEAN summary function and DTYPE and ANL01FL constant variables.
Line 7: Same dataset POSTM as outer SELECT.
Line 8: GROUP BY USUBJID, PARAMN, and AVISITN applies to how MEAN of AVAL is calculated.
Line 9: HAVING selects which record is returned to the outer SELECT based on the GROUP BY variables. In this example, for each grouping of USUBJID, PARAMN, and AVISITN, the minimum value of AWTDIFF is selected, which is the difference between the actual study day and the target study day. By selecting the MIN value of AWTDIFF, the closest study day value to the target study day is selected and returned to the outer SELECT. The assumption is that there may be multiple records per USUBJID, PARAMN, and AVISITN.
Line 11: The records returned from the subquery are joined with the outer SELECT. Because of the RIGHT JOIN in **line 4**, only the matching records are kept, which will represent the newly created AVERAGE records.

1.2.6.6 Option for using one NUM column single result subquery in HAVING/WHERE

```
1. select sex, weight from sashelp.class group by sex
2.    having weight >     /* subquery returns one value */
3.    (select avg(weight) from college.class);
```

Line 1: Choose SEX and WEIGHT columns from SASHELP.CLASS table group by SEX based on WEIGHT greater than the overall average weight from another table, COLLEGE.CLASS. Note that the WARNING message can be ignored since the ORDER BY clause causes an ERROR.
Line 2: Use HAVING clause to apply condition. Can use WHERE clause without GROUP BY.
Line 3: Subquery using SELECT within () is first applied to identify rows — single or multiple row values can be returned (this example returns the average weight value from COLLEGE.CLASS table). Note that ';' is excluded from subqueries. See SAS paper on introduction to subqueries. See SAS paper on subquery examples [11].

1.2.6.7 SAS output

```
Sex        Weight

F             112
F           102.5
F           112.5
M             128
M           102.5
M             150
M           112.5
M             112
M             133
```

1.2.6.8 Option for using one CHAR column single result subquery in HAVING/WHERE

```
1. select sex, weight from sashelp.class group by sex
2.  having sex in          /* subquery returns one value */
3.  (select sex from college.class where sex = 'F');
```

Line 1: Choose SEX and WEIGHT columns from SASHELP.CLASS table group by SEX based on SEX value of 'F' from another table COLLEGE.CLASS. Note that the WARNING message can be ignored since the ORDER BY clause causes an ERROR.
Line 2: Use HAVING clause to apply condition. Can use WHERE clause without GROUP BY.
Line 3: Subquery using SELECT within () is first applied to identify rows — single or multiple row values can be returned (this example returns the 'F' from COLLEGE.CLASS table). Note that SAS automatically quotes character values as part of the subquery process and that ';' is excluded from subqueries.

1.2.6.9 SAS output

```
Sex        Weight

F            50.5
F           112.5
F           102.5
```

```
Sex        Weight

F              90
F             112
F              84
F              98
F            84.5
F              77
```

1.2.6.10 Option for using one-column multiple result subquery in HAVING/WHERE

```
1. select name, sex, age from sashelp.class group by sex
2.     having age ~in          /* subquery can return multiple values */
3.     (select distinct age from college.class where sex = 'F');
```

Line 1: Choose NAME, SEX, and AGE columns from SASHELP.CLASS table group by SEX based on nonmatching female ages from another table COLLEGE.CLASS. Note that the WARNING message can be ignored, since the ORDER BY clause causes an ERROR.
Line 2: Use HAVING clause to apply condition. Can use WHERE clause without GROUP BY.
Line 3: Subquery using SELECT within () is first applied to identify rows — single or multiple rows can be returned (this example returns unique AGE values separated by spaces for females in the COLLEGE.CLASS table).

1.2.6.11 SAS output

```
Name      Sex    Age

Philip    M      16
```

1.2.6.12 Option for using one-column single result subquery in SELECT

```
1. select sex, weight -
2.  (select avg(weight) from college.class)/*subquery returns one value*/
3.     as diff
4.     from sashelp.class;
```

Line 1: Choose SEX and WEIGHT minus the AVG(WEIGHT) from COLLEGE.CLASS.
Line 2: Subquery returns one value, which is used as a constant value in the SELECT clause.
Line 3: Save new value as DIFF.
Line 4: From SASHELP.CLASS table.

1.2.6.13 SAS output

Sex	diff
F	-49.5263
F	-10.0263
F	-23.0263
F	11.97368
M	49.97368
M	27.97368
M	32.97368
M	-15.0263
M	11.97368

1.2.6.14 Option for using multiple columns with multiple results subquery in FROM

```
1. select sex, weight, avgwt from /* select from intermediate table */
2.  (select sex, weight, avg(weight) as avgwt from college.class)
3.   where weight > avgwt;
```

Line 1: Choose SEX, WEIGHT, and AVGWT from the subquery source intermediate table. Note that only the variables from the subquery intermediate table can be selected in the outer SELECT clause. Note also that creating an intermediate table can also be used in inner joins with WHERE and outer joins with LEFT JOIN, for example, to dynamically create a table before joining tables.
Line 2: First choose SEX, WEIGHT, and average of WEIGHT from COLLEGE. CLASS. This is the only table reference. Almost any valid SELECT statement is acceptable. This technique is also referred as an in-line view.
Line 3: Second subset based on condition of WEIGHT greater than AVGWT.

1.2.6.15 SAS output

Sex	Weight	avgwt
M	112.5	100.0263
F	102.5	100.0263
M	102.5	100.0263
F	112.5	100.0263
F	112	100.0263
M	150	100.0263
M	128	100.0263
M	133	100.0263
M	112	100.0263

1.2.6.16 Option for using multiple columns with multiple results subquery in JOIN

```
1. create table mytable as
2. select class.name, students.sex
```

```
3. from sashelp.class as class         /* code for joining 2 tables */
4. left join                           /* left, full, or right join */
5. (select name, sex from mylib.students) as students
6. on class.name = students.name and class.sex = students.sex;

7. create table mytable2 as
8. select class.name, tests.score
9. from sashelp.class as class   /* code for joining 2 tables */
10. right join                    /* right join assures only matching subquery condition */
11. (select name, sex, score from tests) as tests     /* obs are selected */
12. on class.name = tests.name and class.sex = tests.sex;
```

Line 2: Second select NAME from the SASHELP.CLASS dataset and SEX from the STUDENTS dataset. While the variables names are the same as those selected in the subquery, this is not required. Any variable can be selected from the SASHELP.CLASS, as well as the subquery dataset, as long as the linking variables are contained in both the outer and subquery datasets.
Line 5: First choose NAME and SEX from MYLIB.STUDENTS. These variables are required, since they are used in the ON as key variables. Almost any valid SELECT statement with or without conditions is acceptable. This technique is also referred as an in-line view.
Line 8: SCORE is selected from the TESTS dataset.
Line 10: Since TESTS is a different dataset and only matching records are required, a RIGHT JOIN is required to only select matching subquery condition observations. Any other JOIN will not work, since it will incorrectly select all records from the SASHELP.CLASS dataset. This effective technique allows you to delete nonmatching records.
Line 11: Subquery is based on TESTS, a separate dataset. As an alternative, the * can be used to select all variables from TESTS dataset. In addition, a separate WHERE clause is another option to subset TESTS dataset.
Line 12: NAME and SEX variables link both SASHELP.CLASS and TESTS datasets.

1.2.6.17 Option for adding baseline flag variable to lab dataset using multiple columns with multiple results subquery in JOIN

See paper for more information [12].

```
1. create table LAB_Flag as select labs.*,
2. ilv.basefl from labs as lb
3. left join
4. (select max(lbdt) as basedt format date9., aval as base, subjid, trtn,
   avisitn, paramcat, paramcd, "Y" as basefl from labs
5. (where = (lbdt < trtsdt))
6. group by subjid, trtn, avisitn, paramcat, paramcd
7. having max(lbdt) = lbdt) as ilv
8. on lb.subjid = ilv.subjid and lb.trtn = ilv.trtn and lb.avisitn = ilv.avisitn and
   lb.paramcat = ilv.paramcat and lb.paramcd = ilv.paramcd and lb.lbdt = ilv.basedt and
   lb.aval = ilv.base;
```

Line 1: SELECT LABS.* keep all variables and records.
Line 2: ILV.BASEFL is the new baseline flag variable.
Line 3: LEFT JOIN keeps all LAB variables and records.
Line 4: Start of the subquery, which is also considered to be an in-line view. SELECT all grouping variables, basedt, base, and basefl. MAX() to identify the most recent visit in case of multiple.
Line 5: End of the subquery. WHERE to select visits before first dose date.
Line 6: GROUP BY key variables to be used in ON clause.
Line 7: HAVING MAX() to select the most recent visit — consistent with MAX() in subquery SELECT. AS ILV intermediate dataset.
Line 8: ON key variables join with LB key variables.

1.2.6.18 Option for adding group descriptive statistics using multiple columns with multiple results subquery in JOIN

```
1.  create table class2 as
2.  select a.*, b.sexn              /* add group count variable */
3.  from sashelp.class as a
4.  left join                       /* keep all records */
5.  (select name, count(name) as sexn   /* create group count variable */
6.  from sashelp.class where sex > ' '/* apply condition as needed */
7.  group by sex            /* group by variable */
8.  ) as b on a.name = b.name;      /* link by detail variable */
    quit;

    data class;
     set sashelp.class;
     if age < 15 then agegrp='Less than 15';
     else if age >= 15 then agegrp='Greater and or equal to 15';
    run;

    proc sql;

    * Repeat for each group variable;
9.  create table class3 as
10. select a.*, b.sexn, c.agen, c.agemx   /* add group count variables */
11. from sashelp.class as a
12. left join
13. (select sex, count(sex) as sexn        /* first subquery */
14. from sashelp.class where sex > ' '  /* two records - male, female */
15. group by sex
16. ) as b on a.sex = b.sex             /* link by summary variable */
17. left join                    /* second subquery */
18. (select name, count(age) as agen, max(age) as agemx/* Count & Max */
19. from sashelp.class where age >.   /* stats on numeric variable */
20. group by agegrp
21. )as c on a.name = c.name          /* link by detail variable */
22. ;
```

Line 1: Remerge group counts — <group_variable>n.
Line 2: SELECT all source variables and SEXN.

Line 4: LEFT JOIN to keep all records.
Lines 5–7: Subquery — count and link by name, group by sex.
Lines 9–21: Repeat for multiple group variables — SEXN and AGEN.
Line 10: Include both count and max or AGE.
Line 18: Calculate both AGEN and AGEMX group statistics.
Line 20: AGEGRP variable groups ages, for example, into three groups. In general, numeric variables will be grouped instead of directly grouping by individual numeric values.

1.2.6.19 SAS output for lines 1–8

Obs	Name	Sex	Age	Height	Weight	sexn
11	Joyce	F	11	51.3	50.5	9
12	Judy	F	14	64.3	90.0	9
13	Louise	F	12	56.3	77.0	9
14	Mary	F	15	66.5	112.0	9
15	Philip	M	16	72.0	150.0	10
16	Robert	M	12	64.8	128.0	10
17	Ronald	M	15	67.0	133.0	10
18	Thomas	M	11	57.5	85.0	10
19	William	M	15	66.5	112.0	10

1.2.6.20 Option for adding group descriptive statistics using multiple columns with multiple results subquery grouped by a different variable in JOIN

```
1. create table sales2 as
2. select a.*, b.salesn, b.salesmn, b.salesmx
3. from sashelp.shoes as a
4. left join
5. (select product, count(sales) as salesn, min(sales) as salesmn, max(sales) as salesmx
6. from sashelp.shoes where sales >.
7. group by product
8. ) as b on a.product = b.product;
```

Line 2: SALES count, min, and max by PRODUCT.
Line 5: calculate count, min, and max of SALES by PRODUCT.

1.2.6.21 SAS output (partial)

Obs	Product	salesn	salesmn	salesmx
1	Boot	52	1179	286497
2	Boot	52	1179	286497
3	Boot	52	1179	286497
4	Boot	52	1179	286497
5	Boot	52	1179	286497
6	Boot	52	1179	286497
7	Boot	52	1179	286497
8	Boot	52	1179	286497
9	Boot	52	1179	286497
10	Boot	52	1179	286497

1.2.6.22 Option for using shorthand version to add group descriptive statistics using multiple columns with multiple results subquery in JOIN

```
1. select *, n(height) as ht_jn, mean(height) as ht_jave from
2.    (select *, n(weight) as wt_agrpn, mean(weight) as wt_agrpave from
3.      (select *, n(age) as age_sgrpn, mean(age) as age_sgrpave
4.    from sashelp.class
5.      group by sex)
6.     group by age)
7.    group by name;
```

Line 1: SELECT *, count of HEIGHT, and MEAN of HEIGHT.
Line 2: First subquery, SELECT *, count of WEIGHT, and MEAN of WEIGHT.
Line 3: Second nested subquery, SELECT *, count of AGE, and MEAN of AGE. Since the nested subquery selects all variables, then all variables are also available in the first outer subquery.
Line 4: From SASHELP.CLASS for nested subquery.
Line 5: GROUP by SEX for nested subquery.
Line 6: GROUP by AGE for outer subquery.
Line 7: GROUP by NAME.

1.2.6.23 SAS output (partial)

Name	Sex	Age	Height	Weight	age_ sgrpn	sgrpave	wt_ agrpn	wt_ agrpave	ht_jn	t_jave
Alfred	M	14	69	112.5	10	13.4	4	101.875	1	69
Alice	F	13	56.5	84	9	13.22222	3	88.66667	1	56.5
Barbara	F	13	65.3	98	9	13.22222	3	88.66667	1	65.3

1.2.6.24 Option for using correlated subqueries

```
1. select prodnum, prodname, prodtype
2. from products
3. where not exists       /* alternative to using EXCEPT operator */
4. (select * from invoice      /* returns matching PRODNUM values */
5. where products.prodnum = invoice.prodnum);/* join with outer table */
```

Line 1: Selects PRODNUM, PRODNAME, and PRODTYPE columns. The objective is to identify products shipped that have not yet been invoiced; i.e., these products exist in the PRODUCTS table, but not in the INVOICE table.
Line 2: PRODUCTS table is accessed by outer query.
Line 3: Nonmatching products are returned by first accessing PRODUCTS.PRODNUM and comparing to INVOICE.PRODNUM to return matching values, which are then excluded in the outer query. EXISTS, NOT EXISTS, ANY, and ALL keywords are acceptable. ANY compares any individual value with each other. ALL compares collection of all values.
Line 4: INVOICE table is accessed by inner query. While SELECT * selects all columns, it may be restricted to a selected column if WHERE included that column name.

Line 5: INVOICE table is linked with PRODUCTS table by PRODNUM. This is the reason this subquery is called correlated.

1.2.6.25 Adding group percentages

Option for Adding Group Percentages Using Results Subquery GROUPed : Goal is to display # Events, # Patients and Patient count percentages per row for each unique DV. (One Patient to Many Events)

```
* 1a - Checklist of all values, most useful to show zeros for no cases in checklist, do
difference with AE terms;
   proc sort data=event nodupkey out=event_list (keep=aeterm
    rename=(aeterm=name));
    by aeterm;
   run;

   * Step 2 - Proc Sort by PT and AETERM;
   proc sort data=event;
    by pt aeterm;
   run;

   * Step 3 - Data Step;
   data master1;
    retain grdtot 0;
    set event;
    by pt aeterm;
    if first.pt then grdtot=grdtot + 1;
    if first.aeterm and aeterm > '' then denom=1;
   run;

   * Step 4 - Proc SQL Step add groups counts;
   proc sql;
   select count(distinct(pt)) into :grdtot from pats;

   create table master2 as
    select unique a.name, b.aeterm, c.pd_catn, c.subjectn
    , left(put(c.pd_catn, 4.) || ' (' || strip(put(c.subjectn, 4.)) || ')')
    as eventlbl1

    , left(put((c.subjectn/&grdtot*100), 4.2) || ' % (' ||
    strip(put(c.subjectn, 4.)) || '/' || strip(put(&grdtot, 4.)) || ')') as
    eventlbl2
    , '0 (0)' as eventlbl3
    , '0% (0/' || strip(put(&grdtot, 4.)) || ')' as eventlbl4
    from event_list as a
    left join master1 as b on a.name=b.aeterm
    left join
     (select unique pt, aeterm, count(aeterm) as pd_catn, count(denom) as
     subjectn
      from master1 group by aeterm
     ) as c on b.pt=c.pt and b.aeterm=c.aeterm ;
  quit;
```

```
* Step 5 - Data Step Adjustment;
data master3;
 set master2;
 varlab=name;

 if aeterm = '' then do;
  eventlbl1=eventlbl3;
  eventlbl2=eventlbl4;
 end;

 label varlab = 'AE Term'
     eventlbl1 = 'Total Events (Patients)' eventlbl2 = ' % (n/N)';
  if varlab > '';
 keep varlab eventlbl1 eventlbl2 pd_catn subjectn;
run;

ods listing;
proc print data=master3;
 var varlab eventlbl1 eventlbl2;
run;

* Step 1a - Create test records if needed;
data master;
 set master;
 if pt ='E201' then do;
  * all pt with this dv;
  dv='XXX';
  output;
 end;
run;

* Step 1b - Create list of all unqiue DV;
proc sort data=master nodupkey out=prodev_list(keep=dv rename=(dv=name));
 by dv;
run;

* Step 2 - Proc Sort by PT and DV;
proc sort data=master;
 by pt dv;
run;

* Step 3 - Data Step;
data master1;
 retain grdtot 0;
 set master;
 by pt dv;
 if first.pt then grdtot=grdtot + 1;
 if first.dv and dv > '' then denom=1;
run;

* Step 4 - Proc SQL Step add groups counts;
proc sql;
select max(grdtot) into :grdtot from master1;
```

```
create table master2 as
select unique a.name, b.dv
 , left(put(c.pd_catn, 4.) || ' (' || strip(put(c.subjectn, 4.)) || ')') as eventlbl1
 , left(put((c.subjectn/&grdtot*100), 4.) || ' % (' || strip(put(c.subjectn, 4.)) || '/' ||
    strip(put(&grdtot, 4.)) || ')') as eventlbl2
 , '0 (0)' as eventlbl3
 , '0% (0/' || strip(put(&grdtot, 4.)) || ')' as eventlbl4
from prodev_list as a
left join master1 as b on a.name=b.dv
left join
(select unique pt, dv, count(dv) as pd_catn, count(denom) as subjectn from master1
   group by dv
) as c on b.pt=c.pt and b.dv=c.dv
;
quit;

* Step 5 - Data Step Adjustment;
data master3;
 set master2;

 varlab=name;
 if dv = '' then do;
  eventlbl1=eventlbl3;
  eventlbl2=eventlbl4;
 end;

 label varlab = 'Protocol Deviation'
     eventlbl1 = 'Total Events (Patients)'
     eventlbl2 = ' % (n/N)';

 if varlab > '';
 keep varlab eventlbl1 eventlbl2;
run;
```

Step 1 - Create test records and dataset of all unique DV values. Can use PROC FORMAT with PUT() in Data Step to create master list of unique DV values.

Step 2 - Proc Sort by PT and DV;

Step 3 - Data Step;
Assign grdtot = 1 for each first patient record, this will be the total number of patients in the study
Assign denom = 1 for each first patient and non-missing dv record;

Step 4 - Proc SQL Step add groups counts;
Create four eventlbl1-4 variables;
eventlbl1 - Total Events (Total Patients with Events);
eventlbl1 - PD_CATN (SUBJECTN);
eventlbl2 - Patient Count with Events/Total Patients % (Patient Count with Events/Total Patients);
eventlbl2 - SUBJECTN/GRDTOT % (SUBJECTN/GRDTOT);
eventlbl3 - 0 (0) constant for no events;
eventlbl4 - 0% (0/Total Patients) constant for no events;

Create macro variable of fixed count;
- A dataset is the master list and is the main dataset;
- B dataset is the data dataset;
- C dataset adds the PD_CATN event counts per patient and the SUBJECTN patient counts per events, they are determined by grouping by events and linking by subject and events;

Step 5 - Data Step Adjustment;
Replace no event records with constants;
eventlbl1 - Total Events (Total Patients with Events);
eventlbl2 - Patient Count with Events/Total Patients % (Patient Count with Events/ Total Patients);
eventlbl3 - 0 (0) constant for no events;
eventlbl4 - 0% (0/Total Patients) constant for no events;

1.2.7 Sort Tables

When sorting tables, you can sort by rows or by groups without having to presort first. As needed, you can apply CALCULATED columns in ORDER BY or GROUP BY clauses. For example, GROUP BY and ORDER BY clauses may be used together with summary functions to group by one column and then sort descriptive statistics by another column. As an alternative to column names, references can be made to SELECT column order number such as 1 and 2.

1.2.7.1 Option for sorting by rows

```
1. order by name, sex DESC        /* DESC is after the column name */
```

Line 1: Sort table by NAME and SEX columns without input table being presorted. Separate multiple columns with comma, ','. Note that functions such as SUBSTR(NAME, 1, 4) may also be used in the ORDER BY clause. Apply DESCending after column name if needed for descending order. Default is ascending.

1.2.7.2 Option for sorting by groups

```
1. group by name, calculated myage
```

Line 1: Group table by NAME and MYAGE columns. Note that MYAGE creation is limited to across column calculation instead of down column calculation. This is often used with summary functions in SELECT clause but not required. It is required for subset conditions using the HAVING clause.

1.2.7.3 Option for sorting by groups and rows

```
1. group by name, calculated age
2. order by name, age, weight
```

Line 1: Group table by NAME and age columns.
Line 2: Within each group, order rows by NAME, AGE, and WEIGHT columns. Note that you can ORDER BY columns that are different from the GROUP BY columns.

1.3 CREATING MACRO VARIABLES

PROC SQL provides several ways to create macro variables. One or more macro variables can be created storing one or more numeric or character values. Useful to apply %PUT statement after to display macro values. Note that only columns used to create macro variables can be listed in the SELECT statement. INTO: is similar to DATA step's SYMPUT() function. See SAS website for more information using PROC SQL with macros. See SAS website for more information on INTO option.

1.3.1 Five Options for Creating Macro Variables

```
1. select presult into :fnpv from mylib.statresult;

2. select presult into :fnpv separated by ',' from mylib.statresult;

3. select left(put(count(patient), 3.))
   into :fnpv1 - :fnpv5        /* :fnpv9999 for unknown */
   from mylib.statresult where tx ne. and study = 94123
   group by tx;

4. select avg(salary), min(salary), max(salary) into
   :mean, :min, :max from mylib.hrstaff;

5. select case
   when round(p_cmhrms, 0.001) < 0.05 then
   (put(p_cmhrms, 5.3) || '*')
   else put(p_cmhrms, 5.3)
end into :fnpv from mylib.statresult;
```

Line 1: Create FNPV macro variable — expect single PRESULT value. If multiple values exist, then only the first value is saved to the macro variable. Applying %LET FNPV = &FNPV; is helpful to remove any leading or trailing blanks. An alternative method is to apply the 'SEPARATED BY' option even if only one value is saved.
Line 2: Create FNPV macro variable — expect multiple PRESULT values separated by ','. Any delimiter character can be specified.
Line 3: Create FNPV1 — FNPV5 macro variables using GROUP BY; expect result as five values based on condition. Can automate with another macro variable from COUNT(TX) or set to :FNPV9999 if number expected is unknown. Max number of 9999 macro variables.
Line 4: Create MEAN, MIN, and MAX macro variables from summary functions. Separate multiple columns and macro variables by commas ','.
Line 5: Create FNPV macro variable from the THEN expression based on the WHEN numeric expression of ROUND(P_CMHRMS) in CASE condition. Note that for macro variables storing character values, the TRIM() and '!!' are useful to remove blanks between characters.

1.4 TABLE STRUCTURE OPERATIONS

Table structure operations that are available with the DATA step are also available with PROC SQL. Columns can be created, altered, or dropped. Note that PROC SQL ignores NUM length specifications less than eight. Best to use DATA STEP to override default of eight. Note also that with these statements, data values are not queried or used for subsetting. See SAS website for more information on referential integrity constraints [13] to improve the quality of data entry (NOT NULL, UNIQUE, CHECK, and PRIMARY KEY).

1.4.1 Five Options for Creating or Modifying Table Structure

```
1. create table mylib.myclass (keep = name age weight) like sashelp.class;

2. create table mylib.newclass       /* variable type = CHAR, DATE, NUM */
   (subject char(20) not null,  /* list of all columns in creation order*/
   birth date label = 'DOB' unique,   /* all attributes after each column*/
   salary num label = 'Salary' length = 7 format = comma10.2);

3. alter table mylib.newclass              /* modify column attributes */
   modify subject label = 'Subject'      /* one clause per column */
   drop salary
   add age num format = 3. label = 'Age';

4. alter table mylib.newclass         /* assure one of limited values */
   add constraint check_sex check(sex in ('M', 'F'));

5. drop table mylib.myclass, mylib.newclass;          /* delete tables */

6. create index name on mylib.students(name);         /* simple index */

7. create index namsex on mylib.students(name, sex); /* composite index */
```

Line 1: Create MYLIB.MYCLASS table by copying all columns and attributes from SASHELP.CLASS table without copying any data rows. Can apply dataset option, such as KEEP to keep NAME, AGE, and WEIGHT columns. See Section 1.5 Table Content Operations.
Line 2: Create new MYLIB.NEWCLASS table with columns and attributes within '()'. SUBJECT column with type and length. Either character, CHAR, or numeric, NUM, columns can specify the length in '()' or the LENGTH = keyword. Note that CHAR or NUM are only applicable with the CREATE TABLE statement. Default length is eight. BIRTH column with label. SALARY column with label and format. Both BIRTH and SALARY columns have default length of eight, even if specified with less than eight. Multiple columns are separated by commas ','. Columns are created in the order specified. For each column, integrity constraints can be added with these keywords — NOT NULL, UNIQUE. Variable types are CHAR, DATE, and NUM. See Section 1.2 Table Access and Retrieval.

Line 3: Change MYLIB.NEWCLASS table structure with one clause for each column. To MODIFY SUBJECT column with label. To DROP SALARY column. To ADD AGE column with attributes (numeric, format, label).
Line 4: Integrity constraint check_sex is created to assure only 'M' or 'F' values are entered in SEX column.
Line 5: Delete MYLIB.MYCLASS and MYLIB.NEWCLASS tables. Multiple tables are separated by commas ','. Permanent or temporary tables (TABLE), views (VIEW), or indexes (INDEX) may be deleted. Note that indexes require a table name.
Line 6: Create a simple index file called NAME on NAME column — generally, a unique key column. The UNIQUE keyword can be added before INDEX to prevent duplicate index values. To have SAS apply indexing before sequential processing, specify (IDXWHERE = YES) as a dataset option. To control indexing, specify (IDXNAME = <index_name>) as a dataset option.
Line 7: Create a composite index file called NAMSEX from NAME and SEX columns. Multiple columns are separated by ','. Note that composite index files cannot have the same name as a simple index file name and that only one index file can be created for a set of columns. Once created, indexes are automatically applied to increase performance. Note that it is not possible to create indexes on views. See SAS paper on indexes [14]. See SAS paper on processing large files [15].

1.4.2 SAS Output for Line 1

```
                 The CONTENTS Procedure
        Alphabetic List of Variables and Attributes
         #        Variable     Type          Len
         2        Age          Num             8
         1        Name         Char            8
         3        Weight       Num             8
```

1.5 TABLE CONTENT OPERATIONS

Table content operations that are available with the DATA step are also available with PROC SQL. Data values can be updated, inserted, or deleted as separate PROC SQL statements. Note that table structure will not change with these statements. For the example below, MYCLASS table has three columns — NAME, AGE, and WEIGHT.

1.5.1 Five Options for Changing Table Content

```
1. update mylib.myclass      /* updates multiple rows */
   set age = age*10,
       weight = weight + 2
   where name in ('Alfred', 'Alice');   /* optional update condition */
```

```
2. insert into mylib.myclass
   values ('Sunil', 40, 150)   /* column names are not specified */
   values ('Joe', 50, 175);    /* match column order - name, age, weight */

3. insert into mylib.myclass
   (age)              /* insert age values */
   values (40)           /* name and weight set to missing */
   values (50);

4. insert into mylib.myclass (name, age)   /* weight is set to missing */
   select name, age from sashelp.class where sex = 'F';

5. delete from mylib.myclass
   where name = ' ';
```

Line 1: Update MYLIB.MYCLASS table. Update AGE column by multiplication by 10. To update WEIGHT column by adding 2. Multiple column expressions are separated by ','. If the condition of NAME is met, optional else updates are made to all rows.
Line 2: Add new rows to MYLIB.MYCLASS table. Because column names are not specified, order of values must match column order. To create new row with values NAME = 'Sunil, numeric AGE = 40 and WEIGHT = 150. To create new row with values NAME = 'Joe', numeric AGE = 50 and WEIGHT = 175. Values are separated by ','. Need place holders, '.' or ' ' for other columns.
Line 3: Add new rows to MYLIB.MYCLASS table. To add following values to AGE column. To create new row with value AGE = 30. To create new row with value AGE = 20. NAME and WEIGHT columns are set to missing. Multiple column names are separated by ','.
Line 4: NAME and AGE values for females from SASHELP.CLASS table are inserted in the MYLIB.MYCLASS table. WEIGHT column is set to missing. Note that multiple columns are separated by ',' and that the order of columns in INSERT and SELECT clauses must match.
Line 5: Delete rows in MYLIB.MYCLASS. Apply NAME condition to select rows for deletion.

1.6 CONNECTING TO RELATIONAL DATABASES

Using the SAS/ACCESS interface, the power of PROC SQL is extended to the relational database by directly selecting columns and applying subsetting conditions to the database with the pass-through facility. Use caution and closely follow the SQL flavor supported by the relational database. For example, SAS-specific SQL options, such as FORMAT =, may not be allowed. The following example shows connection to ORACLE database. See intro SAS paper on SQL pass-through [16]. See SAS paper using lib names to optimize SQL queries [17]. See SAS paper using ODBC and SAS/Access [18]. See reference to SAS/Access [19]. See reference to SAS/Access by Host [20].

1.6.1 Basic Lines of Code for Using LIBNAME to Access ORACLE Relational Database

```
1. libname billing oracle
2. user = 'sgupta' dpprompt = YES
3. path = "financials" schema = claims;
```

Lines 1–3: After the LIBNAME statement, BILLING defines the connection to CLAIMS schema as identified by the network short path name financials. Once the BILLING libname is defined, it can be used to access the ORACLE databases such as BILLING.CLAIMS_DETAILS. See SAS paper for more information [21].

1.6.2 Basic Lines of Code for SQL Pass-Through to ORACLE Relational Database

```
1. proc sql;

2. connect to oracle as mylink        /* create symbolic link to oracle */
3. (user = me orapw = password path = "@mypath");
4. %put &sqlxrc &sqlxmsg;              /* confirm connection */

5. create table rdbtab as
6. select * from connection to mylink   /* run SQL pass-through code */
7. (select name, sex        /* Only name and sex are selected */
8. from newclass where age > 40);   /* Only age > 40 rows are selected */
9. %put &sqlxrc &sqlxmsg;

10. disconnect from mylink;            /* delete symbolic link */

11. quit;
```

Line 1: PROC SQL statement is required.
Line 2: CONNECT with ORACLE server using MYLINK temporary name. CONNECT TO are required keywords. ORACLE is one of the available databases that can be accessed along with DB2, SYBASE, etc. Use ACCESS instead of ORACLE to access Microsoft Access databases.
Line 3: Any required information to access the table such as your user ID, password, and directory path can be specified. For accessing Microsoft Access databases, use path = "C:\projecta\myaccess_data.mdb".
Line 4: &SQLXRC and &SQLXMSG are useful automatic macro variables to confirm connection. See automatic macrovariable section. See SAS website for more information on automatic macrovariables [22].
Line 5: Save results in a table instead of just displaying the query results.
Line 6: General syntax to select all columns from the server connection MYLINK. CONNECTION TO keywords are required.
Line 7: Apply SQL pass-through facility to select NAME and SEX columns. Note that all database host functions are available.
Line 8: Apply AGE condition directly on NEWCLASS table within oracle database.
Line 9: Confirm rows were retrieved.
Line 10: DISCONNECT MYLINK when complete.
Line 11: Quit statement is required.

1.6.3 Basic Lines of Code for SQL Pass-Through to DB2 Relational Database

```
1. proc sql;

2. connect to db2 (ssid = mydb2);     /* create symbolic link to mydb2 */
3. execute                           /* run following SQL pass-through code */
4. (select name, sex                 /* Only name and sex are selected */
5. from newclass where age > 40)     /* Only age > 40 rows are selected */
6. by mydb2;
7. %put &sqlxrc &sqlxmsg;

8. disconnect from mydb2;            /* delete symbolic link */

9. quit;
```

Line 1: PROC SQL statement is required.
Line 2: CONNECT with DB2 server using MYDB2 temporary name. CONNECT TO keywords are required.
Line 3: EXECUTE keyword is required to run **Lines 4** and **5**.
Lines 4–5: To apply SQL pass-through facility to select NAME and SEX columns.
Line 6: Sort rows by mydb2 table.
Line 7: Confirm rows were retrieved.
Line 8: DISCONNECT MYDB2 when complete.
Line 9: QUIT statement is required.

1.6.4 Basic Lines of Code for SQL Pass-Through to Database Using ODBC

```
1. proc sql;

2. connect to odbc (dsn = mydb2 uid = me pwd = password); /* create link */
3. create table rdbtab as
4. select * from k.org               /* Select all columns */
5. for fetch only);

6. disconnect from odbc;             /* delete symbolic link */

7. quit;
```

Line 1: PROC SQL statement is required.
Line 2: CONNECT with DB2 server using ODBC temporary name. CONNECT TO keywords are required.
Line 3: Create new table RBDTAB.
Lines 4–5: To apply SQL pass-through facility to select all columns.
Line 6: DISCONNECT ODBC when complete.
Line 7: QUIT statement is required.

See the following tables for quick reference to applying PROC SQL.

TABLE 1.1 Comparing PROC SQL with DATA step

DATA STEP (SAS PROGRAMMING)	*PROC SQL (DATABASE PROGRAMMING)*
Dataset, observations, variables	Tables, rows, columns
SAS functions with COALESCE() — multiple variables for across the record processing	SAS functions with COALESCE() — multiple variables for across-the-record processing and single variable for down-the-record processing
Dataset options	Dataset options
IF-THEN-DO-END, CASE-SELECT-ELSE Statements	CASE-SELECT-ELSE Clause
Do Loop, Output	One-to-Many join can simulate Do Loop
Space to separate variables	Comma ',' to separate variables
New variable = valid expression;	Valid expression AS new variable
IF/WHERE Statements	WHERE for details/HAVING for summaries
BY DESCENDING AGE;	ORDER BY AGE DESCENDING
Multiple SAS statements	Only one SAS statement
By default, includes all variables	By default, excludes all variables
Many-to-many merge	Cartesian Product is better to control the type of variable date range joining by a common key variable
By default, If A or B; Full Outer	By default, If A and B; Inner Join
If A; If B; If A and B; If A or B; If A not B;	Joins: Left, Right, Inner, Full, Except
Set A B;	Sets: Outer Union Corr
Can recycle dataset and variable names	Requires new dataset and column names
NA	Unique PROC SQL keywords

Note: For more information, see SAS paper for DATA step diehards [23]. See SAS paper on top 10 reasons for using PROC SQL [24]. See SAS paper on comparing program efficiency [25]. See SAS paper on why PROC SQL is a must-know skill [26].

TABLE 1.2 No PROC SQL equivalent

DATA STEP ONLY	*DESCRIPTION*
FIRST (DOT), LAST (DOT)	By Group Processing using IF FIRST (DOT)
OUTPUT;	Write record to dataset
LAG()	Access values from previous record
ARRAY VAR(3) $ VAR1 – VAR3;	Create and process arrays
DO I = 1 TO 10;	Create and process do-loops

Note: Advanced PROC SQL programming techniques are being developed to simulate IF FIRST (DOT) to subset datasets.

TABLE 1.3 Comparing PROC SQL with SAS procedures

PROC FREQ, PROC MEANS, PROC PRINT, PROC SORT, PROC DATASETS	*PROC SQL*
Proc Freq data = X; tables sex/list; run;	Proc sql; Select distinct sex from X; quit;
Proc Means data = X mean min; var weight; run;	Proc sql; Select mean(weight), min(weight) from X; quit;
Proc Print data = X; var name sex; run;	Proc sql; Select name, sex from X; quit;
Proc Sort data = X; by name; run;	Proc sql; Select * from X order by name; quit;
Proc Datasets; delete X; run;	Proc sql; Drop X; quit;

Note: See SAS paper on comparing PROC SQL with other SAS procedures [27].

TABLE 1.4 Efficiency gains with PROC SQL

TASK	*EFFICIENCY GAINS*
Efficiency Categories	CPU Time, Memory, I/O, Data Storage, Programming Time
Sort large unsorted datasets	Combination of Proc Sort and Data Step
Using presorted large datasets when joining datasets by uncommon variables	(SORTEDBY =) option takes advantage of presorted datasets when joining datasets. Instead of remaining the variables, PROC SQL can join by different variable names.
Simple Index	Logical reference without physically sorting — best if WHERE results in < = 15% of the population group

Note: See SAS paper on comparing DATA step merges with PROC SQL joins [28].

TABLE 1.5 Unique PROC SQL keywords

KEYWORD	*DESCRIPTION*
AS	Creating new columns Ex. ((weight/sum(weight))*100) as wpercent
CALCULATED	Referencing new columns after being specified Ex. where calculated wgroup = 'high'
DISTINCT/UNIQUE	Displaying unique combination of variables Ex. distinct patno
INTO:	Creating macro variables Ex. sum(weight) into :wsum
CHAR/DATE/NUM	Variable type when creating variables Ex. create table mydata (client char(25) format = $25. label = 'Client');

Note: ALL is keyword that is also valid with the DATA step.

TABLE 1.6 Selected useful PROC SQL functions

	SAS FUNCTION DESCRIPTION
Character Function	
INPUT()	Convert character values to numeric values
LEFT()	Shifts and removes leading blanks
PUT()	Convert numeric values to character values
STRIP()	LEFT(TRIM()) removes trailing blanks and shifts and removes leading blanks
TRIM()	Removes trailing blanks
Numeric Function	
AVG(), MEAN()	Calculates the average value of one or more columns
COUNT(), N()	Counts the number of nonmissing values for a column
INT()	Calculates the integer value of the numeric expression
MAX()	Identifies the maximum value from one or more columns
MIN()	Identifies the minimum value from one or more columns
NMISS()	Counts the number of missing columns
RANGE()	Calculates the difference between the smallest and largest values
STD()	Calculates the standard deviation from one or more columns
SUM()	Calculates the sum from a one or more columns
VAR()	Calculates the variance from one or more columns
Date Function	
TODAY()	Returns the current date as a SAS date value
MONTH()	Returns the month from the SAS date variable
INTCK('WEEK',,)	Returns the number of weeks between two SAS date variables

TABLE 1.7 Useful PROC SQL options

KEYWORD	*DESCRIPTION*
_METHOD	Displays PROC SQL execution options
_TREE	Displays visual structure of logic
FEEDBACK	Displays code executed
FLOW	Wraps text within cell
INOBS =	Controls the number of rows read
NOEXEC	Check syntax without executing the code
NOWARN	Prevents displaying WARNING messages
NUMBER	Displays the row number of SELECT clause
OUTOBS =	Controls the number of rows written

TABLE 1.8 Selected automatic PROC SQL macro variables

MACRO VARIABLE	*DESCRIPTION FROM THE MOST RECENT PROC SQL STATEMENT*
&SQLEXISTCODE	Contains the highest number of return code status from insert failures
&SQLOBS	Contains the number of rows selected
&SQLOOPS	Contains the number of PROC SQL inner loop iterations
&SQLRC	Contains the recent return status value (0, 4, 8, 12, 16, 24, 28)
&SQLXMSG	Contains the error code and message from the DBMS connection
&SQLXRC	Contains the return status value from the DBMS connection

TABLE 1.9 PROC SQL dictionary tables

DICTIONARY.<TABLE>, SASHELP.<TABLE>	*KEY VARIABLES*
.DICTIONARIES,.VDCTNRY	memname, name
.MEMBERS,.VMEMBER	libname, path, memname
.TABLES,.VTABLE	libname, memname, nvars, nobs, crdate, modate
.COLUMNS,.VCOLUMNS	libname, memname, name, type, length, label, format

Note: Comparable to PROC CONTENTS and PROC DATASETS. See SAS site for information and webinar on accessing SAS dictionary tables [29].

TABLE 1.10 General PROC SQL usage

PROC sql; create table mylib.newclass as	When defining columns, you can select columns that already exist in the tables or create new columns. Expressions and summary functions could also be used. Column attributes can also be defined.
select name, sex label = 'Sex' from sashelp.class	Alias can be used to reference tables. Joining tables is easy to accomplish with PROC SQL. When joining tables, options include inner or outer joins. Inner joins return a table containing rows that match both tables. Outer joins return a table containing rows that match both tables plus all nonmatching rows form the left, the right, or both tables.
where sex = 'F' order by name;	When subsetting tables using existing or new columns, you can subset by rows or by groups. The difference is in the selection condition. Subsetting by row compares rows to an expression, and subsetting by group compares rows to a group expression, which generally contains a summary function.
quit;	When sorting tables, you can sort by rows or by groups. Sorting by group requires the group by clause.

TABLE 1.11 Comparing PROC SQL join with DATA step merge

DATASET	*MERGE; BY;*	*PROC SQL;*
ALLAB	No conditions	Full outer join
ALLA	If A;	Left outer join
ALLB	If B;	Right outer join
AANDB	If A and B;	Inner join
ANOTB	If A and not B;	Except
BNOTA	If B and not A;	Except

Ultimate PROC SQL Example 1

```
proc sql;
create table base2 as                 /*Add suffix number to base dataset*/
select unique a.*, b.aval, b.dtype    /*Keep all A.* & new B.vars*/
from base (drop = aval) as a          /*Drop AVAL since replaced*/
right join                            /*Save AVERAGE records only*/
(                                     /*Subquery start*/
select unique usubjid, paramn, ady,   /*Keep linking variables*/
mean(aval) as aval,                   /*Summarize to get one key*/
'AVERAGE' as dtype                    /*Flag data type method*/
from base                             /*FROM same dataset as outer*/
```

```
where ady <= 1                  /*Subset only baseline records*/
group by usubjid, paramn        /*GROUP BY key variables for MEAN()*/
having max(ady) = ady           /*Of the records, keep only max ADY*/
)                               /* Subquery stop*/
as b on a.usubjid = b.usubjid and a.paramn = b.paramn and a.ady = b.ady   /*Linking
variables with outer dataset*/
;
quit;
```

SEQUENCE: (SUBQUERY/OUTER WHERE), GROUP BY, MEAN(), HAVING & JOIN
    SUBQUERY Options: SELECT, FROM/JOIN and WHERE/HAVING

The subquery dataset contains only baseline records, WHERE ADY < = 1. ADY is the study day, which is based on the first dose date. This example shows how to get group descriptive statistics by summarizing AVAL by USUBJID and PARAMN. The record that is selected to match with the outer dataset is the maximum value of ADY. Practical application is to summarize and identify the record that is closest to the baseline record, which is MAX(ADY). For example, if there are both Screening and Day 1 records, then select closest to 1. UNIQUE and DROP are applied to assure only one record per USUBJID, PARAMN, and ADY.

In general, the BASE2 dataset is expected to be appended to the corresponding POST2 dataset to store all AVERAGED records. Or an alternative is to append BASE2 and POST2 with BASE and POST to keep both the original records and the AVERAGED records. The DTYPE = 'AVERAGE' indicates the data type source.

Note that this technique is also useful for identifying duplicate records by creating DUPCNT as COUNT(USUBJID). Notice that the HAVING clause variables may be different from the GROUP BY variables. The HAVING clause is applied within the GROUP BY clause variables.

Changing to LEFT JOIN keeps all BASE records. Along with removing the HAVING clause, removing DROPing AVAL and changing AVAL to AVALB adds the AVALB to BASE2 dataset. Detail level subset with the WHERE clause can be applied in the SUBQUERY or outer SELECT as needed. Additionally, LEFT JOIN and SUBQUERIES can be used to add more variables.

Ultimate PROC SQL Example 2

```
proc sql;
/*-- first select all ex variables and then select only for first dose date------------- */

create table ex0 as
select *, min(exstdtn) as exdt
    from odat.ex
    where exstdtn ne.
    group by subject;

delete from ex0 where exstdtn ne exdt;
```

```
*------dose date & visit date (Screening)-------;
* 1 - right join subquery - ex. ex first dose date right join on subject with dov dates,
subset dov.instancename = 'screening';
* 2 - subquery result - ex. temp dataset of ex first dose date and dov screen date;
* 3 - outer query result - ex. final dataset mh, subset mhyn = 'Yes', left join on
subject of temp dataset;
* 4 - final dataset is mh with first dose date and dov screen dates linked by subject;

create table mh0 as
select m.*,
    x.exdt,
    year(x.exdt) as exdt_yy,
    month(x.exdt) as exdt_mm,
    day(x.exdt) as exdt_dd,
    x.visdtn,
    x.visdtn_yy,
    x.visdtn_mm,
    x.visdtn_dd
from odat.mh(where = (mhyn = 'Yes')) m
    left join
    (select datepart(e.exstdtn) as exdt,
        d.visdtn,
        d.visdtn_yy,
        d.visdtn_mm,
        d.visdtn_dd,
        e.subject
    from ex0 e
        right join        odat.dov(where = (upcase(strip(instancename)) = 'SCREENING')) d
        on e.subject = d.subject) x
    on m.subject = x.subject
    order by subject, mhterm, mhstdtn;
quit;
```

REFERENCES

1. The SQL Procedure. http://support.sas.com/documentation/cdl/en/proc/61895/HTML/default/viewer.htm#a000086336.htm.
2. Lafler, Kirk Paul, Undocumented and Hard-to-find SQL Features, SUGI 28. http://www2.sas.com/proceedings/sugi28/019-28.pdf.
3. Cheng, Wei, Helpful Undocumented Features in SAS, SUGI 29. http://www2.sas.com/proceedings/sugi29/040-29.pdf.
4. Overview of the SELECT Statement. http://support.sas.com/documentation/cdl/en/sqlproc/62086/HTML/default/viewer.htm#a001339992.htm.
5. Using the PROC SQL Automatic Macro Variables. http://support.sas.com/documentation/cdl/en/sqlproc/62086/HTML/default/viewer.htm#a001360983.htm#a001404680.
6. SAS 9.1.3 Language Reference: Concepts. http://support.sas.com/documentation/onlinedoc/91pdf/sasdoc_913/base_lrconcept_9196.pdf.

7. PROC SQL Summary Functions. http://support.sas.com/kb/25/279.html.
8. Gupta, Sunil, Something for nothing? Adding group descriptive statistics. http://blogs.sas.com/content/sgf/2012/06/20/adding-group-descriptive-statistics/.
9. Chapman, Tasha, Proc SQL. http://www.sascommunity.org/mwiki/images/b/ba/Proc_SQL.pdf.
10. Cheng, Wei, Helpful Undocumented Features in SAS, SUGI 29. http://www2.sas.com/proceedings/sugi29/040-29.pdf.
11. Zhang, Lei, Danbo Yi, WORKING WITH SUBQUERY IN THE SQL PROCEDURE. http://www.nesug.org/Proceedings/nesug98/dbas/p005.pdf.
12. Vemuri, Pavan, SQL SUBQUERIES: Usage in Clinical Programming, PharmaSUG 2013. http://www.pharmasug.org/proceedings/2013/PO/PharmaSUG-2013-PO15.pdf.
13. Gupta, Sunil, Getting Familiar with SAS® Version 8.2 and 9.0 Enhancements, WUSS 2003. http://www.lexjansen.com/wuss/2003/Tutorials/i-getting_familiar_with_sas_version_8_and_9.pdf.
14. Lafler, Kirk, Efficiency Techniques for Beginning PROC SQL Users, SUGI 29. http://www2.sas.com/proceedings/sugi29/127-29.pdf.
15. Wilcox, Andrew, Efficiency Techniques for Accessing Large Data Files, SUGI 25. http://www2.sas.com/proceedings/sugi25/25/dw/25p115.pdf.
16. Winn, Thomas, Introduction PROC SQL to Using, SUGI 22. http://www2.sas.com/proceedings/sugi22/BEGTUTOR/PAPER67.pdf.
17. Levin, Fred, Using the SAS/ACCESS Libname Technology to Get Improvements in Performance and Optimizations in SAS/SQL Queries, SUGI 26. http://www2.sas.com/proceedings/sugi26/p110-26.pdf.
18. Li, Leiming, A Process for Automatically Retrieving Database Using ODBC and SAS/ACCESS SQL Procedure Pass-Through Facility, SUGI 24. http://www2.sas.com/proceedings/sugi24/Coders/p089-24.pdf.
19. SAS/ACCESS Software. http://www.sas.com/en_us/software/data-management/access.html.
20. SAS/ACCESS Features by Host. http://support.sas.com/documentation/cdl/en/acreldb/63647/HTML/default/viewer.htm#app3-accgen8.htm.
21. Schacherer, Christopher, Michelle A. Detry, PROC SQL: From SELECT to Pass-Through SQL. http://www.scsug.org/SCSUGProceedings/2010/Schacherer_2/PROC_SQL_%20From_SELECT_to_Pass-Through_SQL.pdf.
22. Using the PROC SQL Automatic Macro Variables. http://support.sas.com/documentation/cdl/en/sqlproc/62086/HTML/default/viewer.htm#a001360983.htm#a001404680.
23. Williams, Christianna, PROC SQL for DATA Step Die-hards, SFG 2008. http://www2.sas.com/proceedings/forum2008/185-2008.pdf.
24. Lafler, Kirk, Kirk's Ten Best PROC SQL Tips and Techniques. http://www.sascommunity.org/mwiki/images/b/ba/Kirk%27s_Ten_Best_PROC_SQL_Tips_and_Techniques_%28Wisconsin_Illinois_SAS_Users_Conference%29.pdf.
25. Bhat, Gajanan, Raj Suligavi, Merging Tables in DATA Step vs. PROC SQL: Convenience and Efficiency Issues, SUGI 26. http://www2.sas.com/proceedings/sugi26/p104-26.pdf.
26. Whitlock, Ian, PROC SQL — Is It a Required Tool for Good SAS® Programming?, SUGI 26. http://www2.sas.com/proceedings/sugi26/p060-26.pdf.
27. Shi, Changhong, Sylvianne Roberge, Application of Some Advanced PROC SQL Features in Clinical Trial Programming, PharmaSUG 2003. http://www.lexjansen.com/pharmasug/2003/datamanagement/dm012.pdf.
28. Droogendyk, Harry, Faisal Dosani, Joining Data: Data Step Merge or SQL?, SGF 2008. http://www2.sas.com/proceedings/forum2008/178-2008.pdf.
29. Accessing SAS System Information by Using DICTIONARY Tables. http://support.sas.com/documentation/cdl/en/sqlproc/62086/HTML/default/viewer.htm#a001385596.htm.

CHAPTER 1: ACCESSING DATA USING SQL—QUESTIONS

1. Creating a non-missing variable from multiple datasets?
2. Creating character string macro variables without any blanks?
3. Creating an empty dataset with minimum code and no uninitialized notes?
4. What technique is useful for identifying and adding baseline lab values in clinical trials?
5. In general, when having nested summary functions, such as SUM(SUM(VOL1, VOL2)*10), with a GROUP BY clause, what is the difference in the two SUM() functions?
6. When applying a condition to subset the first dataset A used in a LEFT JOIN with a second dataset B, does PROC SQL subset based on the condition after the join or apply the LEFT JOIN to keep all records in dataset A?
7. Is it possible to select all variables in a dataset as well as apply the COALESCE() function to keep nonmissing values in the same SELECT statement when joining two tables?
8. What PROC SQL option can be added to prevent any warnings?
9. Is it possible to control and loop through each record in a dataset as done with the _N_ in the DATA step?
10. In general, do variables listed in the GROUP BY clause need to be included in the SELECT clause?
11. When submitting PROC SQL code in batch mode, is there a method to submit a line of syntax greater than 256 characters?
12. Is it possible to select more variables when applying EXCEPT or INTERSECT joins?
13. What are some techniques to prevent this 'NOTE: The query requires remerging summary statistics back with the original data'?
14. Is it possible to calculate event percentages, such as total number of patients with events/total number of patients using PROC SQL?
15. What is one advantage that PROC SQL has over DATA step when joining datasets?
16. What is the syntax to apply the colon modifier ':' to match based on values only and exclude blanks?
17. What is the PROC SQL syntax to create macro variables?
18. What is the danger of doing a many-to-many join without a WHERE clause?
19. Is there a method to pull data from multiple datasets and create flag variables for non-missing values to get patient accountability?

SAS Macro Processing

2

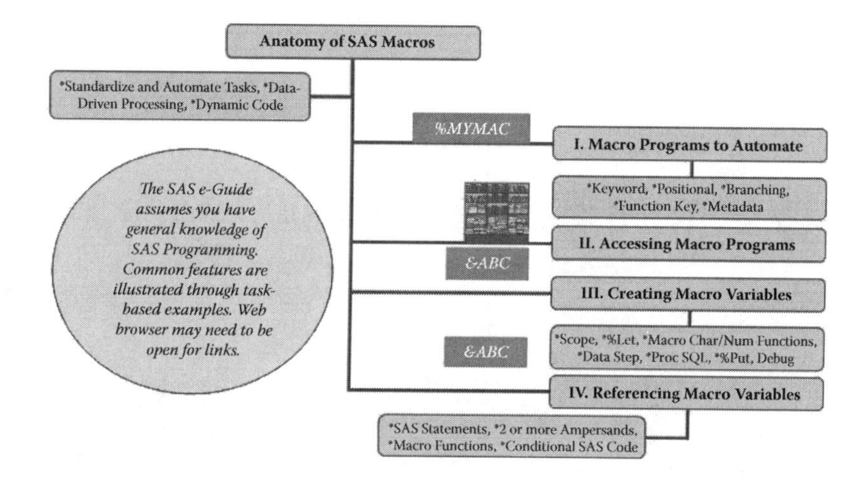

Chapter Overview

2.1 INTRODUCTION

This chapter is organized to facilitate easy searching of key points by summarizing and differentiating the syntax between similar SAS statements and options. The basic syntax for SAS macro processing will be explained and illustrated with simple task-oriented examples. Questions will be included at the end of each section to reinforce the user's knowledge of the topic.

At the beginning of the chapter, a chapter overview is provided to facilitate quick reference to the detailed examples and syntax in the chapter. The basic syntax, expected data, and descriptions are organized in summary tables to facilitate memory recall of the information. General rules within each section list common points about similar statements or options. Each topic includes the basic syntax, a series of key points, and other notes about the specific SAS statement or option. Examples of SAS program and code statements are line numbered with references for more detailed explanation. Note that SAS examples are a complete block of code that can be executed, while selected SAS code syntax needs to be executed as part of the remaining program. This unique approach empowers both the advanced programmer who needs a quick refresher, as well as programmers interested in learning new programming techniques.

2.2 INTRODUCTION TO SAS MACRO LANGUAGE AND GROUND RULES

SAS Macro Processing

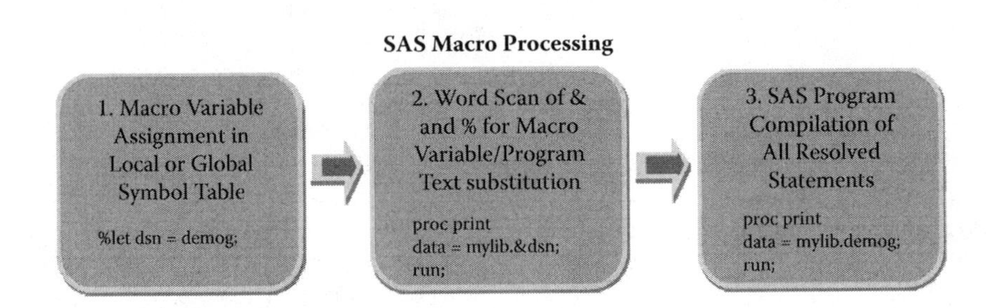

2.2.1 What Is the SAS Macro Language?

The macro language serves as a dynamic character editor for SAS programs. Macros can be used almost anywhere in the SAS program to standardize and automate repeated code, as well as communicate, customize, and create SAS code. Using macros saves retyping time-repeated macro calls. Data-driven processing and dynamic code generation allow for great flexibility. Advanced macro programming technique allows for writing macro programs that call other macro programs or create new macro programs.

2.2.2 General Rules

Macro variables can be defined in open code or in macro programs (%MACRO; %MEND;). Macro statements (%) must be within a macro program and not as open code. Macro variables and macro statements are independent of the Data Step and can be placed anywhere in the program. Macro functions can be used inside or outside the macro program. All input to the macro language is a character string.

The ampersand (&) symbol is used to reference macro variable, and the percent sign (%) symbol is used to reference macro programs.

The macro language system or Macro Facility involves preprocessing or scanning the SAS code before compiling and executing the SAS program. The main purpose of the macro processor is to substitute text from the values stored in the global or local symbol tables.

2.2.2.1 General syntax to compile macro programs

[MYMAC.SAS]

```
1. libname mydata 'c:\mydata';
2. libname compmacr 'c:\mymacros';
3. options mstored sasmstore = compmacr;
4. %macro mymac(libn, dsn =)/store;
5. proc print data = &libn..&dsn;run;
6. %mend mymac;
7. %mymac(mydata, dsn = demog);
```

Line 3: MSTORED and SASMSTORE options are required to save compiled programs in the SASMACR catalog in the COMPMACR library. Both options are also required for usage.
Line 4: %MACRO MYMAC creates the macro program MYMAC. First position parameter, LIBN, is followed by the keyword parameter, DSN. The STORE option directs the compiled macro to the COMPMACR libname as a precompiled version.
Line 5: All code including Data Step and SAS procedures steps between these two statements are within the macro program. In addition, calls to other macros or new local macro variables can be defined. Once the macro is defined, a macro call can be made.
Line 6: %MEND MYMAC ends the macro program. While adding MYMAC is not required, it is generally best practice to know when macro program ends.
Line 7: Executes MYMAC macro as PROC PRINT DATA = MYDATA.DEMOG; RUN; Note that while ';' is not required at the end, it is recommended.

It is strongly recommended to first have a working version of the SAS program before converting it to a macro program and creating macro variables. Note that the name of the SAS program, MYMAC.SAS, needs to be the same name as the macro program MYMAC for the SASAUTOS = option to work.

2.3 WRITING MACRO PROGRAMS TO AUTOMATE TASKS

Macro program names follow the same rules as those for macro variables: 32 characters or fewer in length; starting with a letter or underscore; and containing only letters, numerals, or underscores. Note that macro definitions also refer to macro programs.

Although macro programs are not required to have any parameters, generally positional or keyword parameters are applied for greater flexibility. Generally, for example, the dataset name, variable name, or subset condition are good candidates for macro parameters, since these may change with macro program call. While you can define almost limitless number of parameters, as a general rule, try not to have more than 10 parameters. Macro programs without any parameters may or may not contain global macro variables. Note that, in general, in batch mode, a SAS statement or a macro call statement cannot exceed 256 characters. To prevent record truncation error, one technique is to save the long macro call in a separate file, set LRECL to a large value, such as 32767, and %INCLUDE the file.

There are two ways to pass parameters to macros — keywords and positional. By default, the parameter variables are declared as local macro variables. Generally, it is recommended to apply keyword parameters for setting default values, additional flexibility in specifying macro variables in any order, and better documentation. As an alternative, global macro variables can be applied directly in macro programs. If the methods of positional and keyword parameters are mixed, the positional parameters must come first.

As a best practice, it is helpful to confirm your assumptions of macro parameter values to prevent ERRORs or WARNINGS. This programming technique is called defensive programming to display user messages or exit the macro with an ABORT statement. Generally, when parameters are defined, it should equal some value instead of being missing. Also, each parameter should be well documented (purpose, required vs. optional parameters, etc.) and tested. See SAS paper on examples of defensive programming [1]. See SAS paper on examples of macro utilities [2].

By default, once defined, SAS stores the complied macro in the WORK.SASMACR catalog.

SAS <program-name> –SYSPARM <values>

For SAS batch execution, the SYSPARM system option can be used to pass macro values at execution time without modifying the program. Within the program, apply &SYSPARM as the macro variable name. Please check the SAS website for complete list of automatic macro variables.

2.3.1 Writing Macro Programs without Any Macro Parameters

```
1. %macro mymac;                    /* fixed macro program */
2. proc print data = mydata.demog; run;
3. %mend mymac;
4. %mymac;
```

Line 1: create MYMAC macro program with no parameters or global macro variables. Useful technique to reduce block of SAS statements to minimum macro call keystrokes.
Line 2: PROC PRINT of MYDATA.DEMOG.
Line 3: End of MYMAC macro program.
Line 4: Execute MYMAC macro program to run the same PROC PRINT code.

2.3.2 Writing Macro Programs Using Global Macro Variables

```
1. %let dsn = demog;              /* global macro variable */
2. %macro mymac;
3. proc print data = mydata.&dsn; run;
4. %mend mymac;
5. %mymac;
```

Line 1: Create global macro variable DSN with DEMOG value.
Line 2: Create MYMAC macro program.
Line 3: PROC PRINT of MYDATA and DSN global macro variable.
Line 4: End of MYMAC macro program.
Line 5: Execute MYMAC macro program.

2.3.3 Writing Macro Programs as Two Options for Using Macro Positional Parameters

```
1. %macro mymac(libn, dsn);     /* need to remember order of parameters */
2. proc print data = &libn..&dsn;run;
3. %mend mymac;
4. %mymac(mydata, demog);
```

Line 1: LIBN and DSN are positional local macro variables. Default values are not possible with positional parameters.
Line 2: PROC PRINT using both macro variables. Macro variables can be used in any order.
Line 3: End of MYMAC macro program.
Line 4: Execute MYMAC macro program. Along with values, macro variables can also be used as parameter values. Note that null values can be passed by using commas as placeholders. Make sure to remember the parameter order.

```
5. %macro mylist(var1, var2, var3);          /* three fixed parameters */
6. proc print data = sashelp.class; var &var1 &var2 &var3; run;
7. %mend mylist;

8. %macro mylist/parmbuff;         /* unlimited number of similar parameters */

* Counts the number of variable names passed in as the parameter;
%let n = %sysfunc(countw(&syspbuff, %str(,)));

9. proc print data = sashelp.class;

                                  /* Loops through each variable */
%do i = 1 %to &n;

                                  /* Pulls off each variable to construct each VAR
                                     statement */
```

```
%let var = %scan(%qsysfunc(compress(%bquote(&syspbuff), %str(%(%)))),&i, %str(,));

var &var;                       /* VAR SEX; VAR WEIGHT; VAR AGE; */

%end;

run;

10. %mend mylist;
* Flexible number of similar variables separated by ',';
11. %mylist(sex, weight, age);
```

Line 8: Add (PARMBUFF) for positional parameter macro calls. Notice that &VAR1, &VAR2 and &VAR3 were not defined as in **Line 5**. With the PARMBUFF option, then a list of similar macro parameter values can be processed instead of a fixed number of macro parameters.
Line 11: The &SYSPBUFF macro parameter contains the value of SEX, WEIGHT, and AGE, which are separated by the delimiter (,). This is a simple example showing greater flexibility of macros. See SAS Example on PARMBUFF [3].

2.3.4 Writing Macro Programs as Three Options for Using Macro Keyword Parameters

```
1. %macro varlist(dsn = ae, patno = 1, listvars = &firstv);
2. proc print data = mydata.&dsn; var &listvars; where patno = &patno; run;
3. %mend varlist;
4. %varlist(dsn = demog, patno = 4, listvars = sex weight);

5. %let firstv = aedt;     /* must create in advance */
6. %varlist();             /* macro call using default parameter values */
```

Line 1: DSN, PATNO, and LISTVARS are keyword local variables. Keyword parameters can be equal to missing, to a default value, or to another macro variable.
Line 2: PROC PRINT using each macro variable in various statements.
Line 3: End of VARLIST macro program.
Line 4: Executes VARLIST macro program. Note that LISTVARS macro variable contains a list of variable names — SEX and WEIGHT. The main benefit over positional parameters is that keyword parameters can be in any order and that all parameters need to be specified.
Line 5: Assigns FIRSTV macro variable to AEDT. Required if planning to use default listvars parameter.
Line 6: Executes VARLIST macro program with default values.

```
7. %macro mylist(dsn = ae, var1 = sex, var2 = weight);  /* three parameters */
8. proc print data = mydata.&dsn; var &var1 &var2; run;
9. %mend mylist;
10. %mylist(dsn = demog, var1 = sex, var2 = weight);
```

```
11. %macro mylist/parmbuff;                    /* unlimited number of
                                                  parameters */
12. proc print data = mydata.&dsn; var &var1 &var2; run;
13. %mend mylist;
14. %mylist(dsn = demog, sex, weight, race);
```

Line 11: As an alternative to **line 7**, where applicable, the PARMBUFF option maybe applied for limitless number of parameters, where the last parameter values are of the same type. See SAS Macro Language Reference for examples [4].
Line 14: With the PARMBUFF option, the last macro parameter VAR2 is repeated as VAR3 for RACE. The &SYSPBUFF macro variable contains the value of DSN = DEMOG, SEX, WEIGHT, RACE. This allows for greater flexibility.

2.3.5 Writing Macro Programs Calling and Branching to More Macro Programs

Although nested macro calls are almost limitless, in general, you may want to limit to four nested macro calls for easier maintenance. In general, macro programs not embedded within other macro programs but called are better designs to create more efficient and robust macro programs. Note that before SAS can execute the chain of all macro calls, SAS must first access all of the macro programs. Also be careful not to reuse macro program parameter names and not to access a local macro variable in another macro program.

```
1. %macro mymac(libn =, dsn =);
2. %mac1(newval =);                  /* first nested level macro call by
                                        MYMAC */
3. %mac2(conval =);                  /* second nested level macro call by
                                        MYMAC */
4. %mend mymac;

5. %macro mac1(newval =);
6. %mac1_1(deepval =);               /* second nested level macro call by
                                        MAC1 */
7. %mend mac1;

8. %findvars(h&study, %source_v(demog));   /* macro program call as parameter */
```

Line 1: Create MYMAC macro program.
Line 2: First nested level macro call to MAC1 macro program.
Line 3: Second nested level macro call to MAC2 macro program.
Line 4: End of MYMAC macro program.
Line 5: Create MAC1 macro program.
Line 6: Second nested level macro call to MAC1_1 macro program.
Line 7: End of MAC1 macro program.
Line 8: As an alternative to nested macro calls, macro programs, such as, %source_v, can be called as a keyword macro parameter.

2.3.6 Writing Macro Programs Called by Function Key

To help reduce keystrokes, for more frequently used SAS macros, macro programs can be assigned to function keys. To assign the %NCNT macro program to a function key, you will need to update the Keys Window.

```
1. %macro ncnt;                /* function key macro */
2. submit                      /* required */
3. 'proc sql; select libname, memname, nobs from dictionary.tables; quit;'
4. %mend ncnt;
```

Line 1: Create NCNT macro program. Generally function key macros do not contain any parameters.
Line 2: SUBMIT is a keyword to execute NCNT macro when function key is pressed.
Line 3: Complete block of SAS statements is enclosed in quotes "".
Line 4: End of NCNT macro program.

2.3.7 Writing Macro Programs Leveraging Metadata Intelligence

```
1. %macro mymac(dsn =);                       /* confirm dataset exists */
2. %if %sysfunc(exist(&dsn)) %then %do;       /* macro function */
3. proc print data = &dsn; run;               /* true evaluation block */
4. %end;
5. %else %do;
6. %put Data Set &dsn does not exist;         /* false evaluation block */
7. %end;
8. %mend mymac;
9. %mymac(dsn = demog);
```

Line 1: Create MYMAC macro program.
Line 2: Conditional logic using %SYSFUNC() macro function and EXIST() function based on existence of dataset. If dataset exists, then execute **line 3**, or else execute **line 6**. %SYSFUNC() allows execution of many SAS functions and can also be used to define the starting or ending values of %DO LOOPS.
Line 3: Statement to execute if **line 2** evaluates to true — PROC PRINT.
Line 4: END statement for true evaluation block.
Line 5: ELSE DO statement to start false evaluation block. This is useful to complete IF-THEN-ELSE logic.
Line 6: Statement to execute if **line 2** evaluates to false — user message to SAS log.
Line 7: END statement for false evaluation block.
Line 8: End of MYMAC macro program.
Line 9: Execute MYMAC macro program.

```
10. %macro macexist(mvname);              /* confirm macro variable exists */
11. proc sql noprint;
```

```
12. select 'YES' into :macvar
13. from dictionary.macros where name = upcase("&mvname");
14. quit;
15. %mend macexist;
```

Line 11: Start PROC SQL code.
Line 12: Save the value 'YES' in the MACVAR local macro variable if the macro variable MVNAME exists. Can use &MACVAR for subsequent processing within the MACEXIST macro program. If the MACVAR macro variable was created with PROC SQL outside of the MACEXIST macro program, then the MACVAR macro variable would be global. As an alternative, within the MACEXIST macro program, the %GLOBAL MACVAR statement before the PROC SQL statement will first make the macro variable global.
Line 13: Access DICTIONARY table MACROS.

An alternative to **lines 10** to **15** is the %SYMEXIST() function. Existence of a dataset can be confirmed before executing SAS code. In general, defensive programming is effective to prevent errors. See SAS website for list of DICTIONARY tables [5]. See SAS paper on applying dictionary tables [6].

2.3.8 Writing Macro Programs Using DATA Step CALL Routines

```
1. %let mycall = dir;
2. %syscall system(mycall);          /* execute DOS command */
```

Line 1: DIR is a DOS command.
Line 2: Execute system level command. Note that '&' is missing from MYCALL macro variable.

See Section 2.5 Creating Macro Variables to Store and Replace Text—Six options for using the DATA step section.

2.3.9 Writing Macro Programs Creating Macro Functions

Rules for creating macro functions:

1. Use all macro statements.
2. Create only local macro variables.
3. Allow macro variable to be resolved to be passed back to the calling macro.

```
1. %macro dsexist(dsn);
2. %if %sysfunc(exist(&dsn)) %then Yes;      /* return 'Yes' if dataset exists */
```

```
3. %else No;                              /* return 'No' if dataset does not exist */
4. %mend dsexist;
5. %let dsn_exist = %dsexist(sashelp.class);
```

Line 2: %IF-%THEN logic is applied to return 'Yes' or 'No' value.
Line 5: Assign value, Yes or No, to DSN_EXIST macro variable. Note that macro functions are similar to regular SAS functions except applied at macro programming level.

2.4 ACCESSING MACROS FROM MACRO LIBRARIES

Using macro programs can be achieved by several ways. Other than the program containing the macro code and calling it for a single use, which stores the macro code in WORK.SASMACR, the easiest method to share macro programs is with the SASAUTOS = system option. Another method is to use the %INCLUDE statement to directly include the macro program. Generally, an AUTOEXEC.SAS file is created with the SASAUTOS option and saved in the same location as the SAS.EXE executable file. See SAS paper on SAS AUTOS option [7]. Any SAS macro program in WORK. SASMACR and duplicated in the autocall, for example, will execute the version in WORK.SASMACR and not the version in the autocall.

Priority Search Order of Macro Programs

2.4.1 Five Options for Accessing Macro Programs from Macro Libraries

```
1. filename othrmac catalog 'C:\othermac';
2. options source mrecall mautosource sasautos = (sasautos, othrmac,
      'C:\mymacros', 'C:\hismacros'); /* access one or more libraries */

3. libname compmacr 'C:\mymacros';        /* compiled stored macros */
4. options mstore sasmstore = compmacr;

5. %include('C:\macros\macro1.sas');      /* directly include macro program */
```

```
6. filename mymacros 'C:\macros';            /* reference macro path */
7. %include mymacros('macro1.sas');

8. filename mymacros 'C:\macros.sas';        /* reference macro program */
9. %include mymacros;
```

Line 1: CATALOG of macro programs can be referenced with the FILENAME statement. Note that for SASAUTOS to locate macro programs, the macro program names must match the actual macro program file name.
Line 2: All macros in library are available. MAUTOSOURCE enables the autocall facility. MRECALL causes the macro processor to search autocall libraries for a member not found in a previous search. Macros are called in the order listed. SASAUTOS is required within the SASAUTOS = option to access SAS supplied macro programs. Note that OTHRMAC library is second priority in the library search.
Lines 3–4: Define COMPMACR libname referencing the SASMACR catalog file containing all compiled macro programs. MSTORE option is used with the SASMSTORE option to point to the catalog file. See Section 2.2.2.1 to create compiled stored macro programs with the STORE option. Make sure to save your source macro programs in a safe place for documentation and maintenance, as well as keep both source and compiled macro programs in sync. Note that only SASMACR catalog file can be created and accessed per full file path name.
Line 5: As an alternative to SASAUTOS option, you can directly include the macro program file with the %INCLUDE statement. Make sure the file containing the macro program exists. With this approach, the macro program names do not need to match the macro program file name.
Lines 6–7: As another alternative to SASAUTOS option, you can create a FILENAME statement to reference a collection of macro programs and then directly include each macro program file. This option is easier to include many macro files.
Lines 8–9: As another alternative to SASAUTOS option, you can create a FILENAME statement to reference one macro program and then directly include that macro program file.

2.5 CREATING MACRO VARIABLES TO STORE AND REPLACE TEXT

In general, all macro variable values are treated as character variables and are stored in memory in a macro symbol table. The macro values can also contain numbers. See 2.3 Writing Macros to Automate Tasks for rules on macro variable names. See SAS paper on creating macro variables [8].

If a macro variable already exists and is being redefined, the new value replaces the original value. Macro variables must be in either the global symbol table or one or more separate local symbol tables that are created for each macro definition. To avoid confusion, it is best not to define in both global and local symbol tables.

If macro variable is defined in both global and local symbol tables, then the local symbol table has higher priority.

It is often useful to define macro variables with '_' prefix to distinguish them from dataset variable names; however, it is possible to keep the same names. In general, keep macro variables as local to prevent interference with other macro variables, such as 'I' from %DO Loops. Note that %SYMDEL statement can be used to delete global macro variables.

While many of the options below create macro variables, it is generally best practice to first declare the scope of all macro variables as global or local.

2.5.1 Scope of Macro Variables (Global vs. Local)

```
1. %global _study _fullsty;          /* access macro variables across programs */

2. %macro varlist(prefix, start, stop); /* local macros - keyword, positional*/
   %do i = &start %to &stop; &prefix&i; %end;
   %mend varlist;
   %global %varlist(sales, 1, 5);    /* global SALES1 - SALES5 */

3. %local name;                      /* internal macro program variables */
```

Line 1: By default, all macro variables created using the various statements, such as %LET, macro functions, Data Step, and PROC SQL, are global unless called within a macro definition. Macro variables can be resolved anywhere within the SAS session once they are defined. Multiple macro variables are separated by blanks. Examples of automatic SAS global macro variables are &SYSDATE9. Note that even if the %GLOBAL statement is within a macro program, the macro variables are still global and can be accessed outside of the macro definition.
Line 2: Macro parameters, positional, or keyword, within macro programs are automatically declared as local macro variables. In general, they can only be resolved within the macro programs or the programs they were created in. As an alternative specifying the macro variables, you can call another macro to indirectly reference macro names. It resolves to %GLOBAL SALES1 SALES2 SALES3 SALES4 SALES5;
Line 3: Local macro variables are defined with the %LOCAL statement inside a macro program. Generally, these macro variables are different from the positional and keyword parameters but are used internally within the macro program.

Note that by default, embedded blanks are included, and leading and trailing blanks are ignored unless macro functions or quotes are applied. %LET statements are compile time statements executed in open code. Macro variable names follow the same general rules as dataset variable naming convention, being up to 32 characters long. Macro variable values can range from 0 to 32k bytes long. Once macro variables are created, they may be updated anytime.

Four Types of Tokens: Literal, Number, Name, Special Characters
'hello', 2001123, city, (*, /, +, –, ', ", %, <, >, &, ;)

2.5.2 Using the %LET Statement

```
1. %let dsn = mylib.mydata;          /* libname.dataset */

2. %let lnam = mylib; %let dnam = mydata; %let dsn = &lnam..&dnam; /* mylib.mydata */

3. %let studyn = 2001123;            /* number */

4. %let visitdt = '01JUN2010'd;      /* date constant */
5. %let kvars = city t&gender &calcvar.r;    /* list of variable names */

6. %let matheq = 1 + 2;              /* numeric expression */

7. %let charn = " hello ";           /* includes spaces and quotes */

8. %let var1 = var2; %let &var1 = &dsn (obs = 10); /*dynamic macro variable name */
        /* called by &&&var2 */
```

Line 1: Creates and assigns value — DSN is the macro variable name, and MYLIB. MYDATA is the constant value. Macro variable names must conform to usual SAS naming conventions. The MYLIB.MYDATA can be literal ('any text – symbol '.' or blank'), number (25), name (item3), or value from 0 to 32k bytes. No quotes are needed around the value, and it can be used to almost include any nonspecial character. Note that character values are case sensitive.
Line 2: Macro variable DSN consists of LNAM and DNAM macro variables. The first '.' is used as a stopping character, and the second '.' is kept to resolve as %LET DSN = MYLIB.MYDATA;. As needed, multiple '.' can be inserted to keep extra '.'.
Line 3: Any number can be stored without quotes.
Line 4: Any date value. An alternative is %LET BASEDATE = 01JUN2010; %LET VISITDT = %SYSEVALF("&BASEDATE"d);
Line 5: Macro variable can be a list of values to process. Three text values are separated by embedded blanks. The &GENDER is a suffix macro variable with 't' attached once resolved. The CALCVAR is a prefix macro variable, itself followed by the stopping character '.'. Once resolved, the suffix 'r' will be attached to the value.
Line 6: Evaluates to 1 + 2.
Line 7: Evaluates to " hello ". Useful to preserve leading and trailing blanks. In general, it is easier to supply quotes than it is to work with quotes as macro values. See next section on quoting functions.
Line 8: Sets macro variable VAR1 equal to VAR2. Evaluates to %LET VAR2 = MYLIB.MYDATA (OBS = 10); when called as &&&VAR2. The general rule is to use the same name when creating the new macro variable name as when calling it after the triple '&&&'.

2.5.3 Using Macro Functions

```
1. %let prnt    = %str(proc print data = &study..demog; run;);
            /* special character values including macro variables */

2. %let prnt    = %nrstr(proc print data = &study..demog; run;);
   %put &prnt;    /* display only - proc print data = &study..demog; run; */
```

```
3. %let onam_cnt =   25 ;/*similar to %let nam_cnt = %trim(%left(&onam_cnt));*/
   %let onam_cnt = %str( 25 );/* onam_cnt stored as (2 blanks)25(2 blanks) */

4. %let nam_cnt = %trim(%left( 25 ));      /* similar to %let nam_cnt = 25; */

5. %let old_val = 2; %let new_val = %eval(&old_val + 2);
   /* similar to %let new_val = 4; NOT %let new_val = 2 + 2; */

6. %let y_n    = yes;
   %let y_n    = %upcase(%substr(&y_n, 1, 1)); /* standardize to 'Y' */

7. %let sqvar = %str(a # b # c);
   %let sqvar = %sysfunc(tranwrd(%sysfunc(compbl(&sqvar)), %str(#), %str(-)));
   %put &sqvar;      /* Reassign sqvar macro variable as a - b - c */
```

Line 1: %STR() macro quoting function preserves leading and trailing blanks. To include the semicolon ';' as part of the string without having SAS act on this token, use the %STR() macro quoting function to mask special characters. This allows creating macros that contain a collection of SAS statements or unbalanced quotes.
Line 2: To include other macro variables (&) or macro calls (%) as part of the string, use the %NRSTR() macro quoting function. %NRSTR() macro quoting function preserves leading and trailing blanks, as well as prevents immediate resolution of any macro variable. Once resolved, the first '.' from &STUDY.. is resolved to leave the study number '.' demog as the libname and dataset name. Often used with %PUT for displaying What You See Is What You Get since the %NRSTR() function is not intended to execute the code.
Lines 3 and **4**: All four examples are similar since leading and trailing blanks are ignored. Use macro functions to preprocess and evaluate the value before assigning it to the macro variable. Macro expressions can be used to create macro variables based on other macro variables. Macro functions work the same way as their corresponding functions. In this example any blanks before and after the number 25 are removed before assigning it to the new macro variable NAM_CNT. To include spaces before and after 25 as the macro value, for example, use %STR(25) with a space before and after 25.

See SAS website for a list of SAS supplied autocall macro functions [9]. See Section 2.6 Referencing Macro Variables to Substitute Text. See Macro Functions in Section 2.6 for more information. See SAS paper on macro functions [10]. See SAS paper on quoting macro functions [11].
Line 5: With the %EVAL() function, the macro variable value 2 is added to the constant 2 and stored as the value 4. Without the %EVAL() macro function, the stored macro value would be '2 + 2'. Use %SYSEVALF() to preprocess continuous values, such as 2.1.
Line 6: Accepts 'YES', 'NO', 'Y', 'N', 'yes', 'no', 'y', or 'n' values for Y_N macro variable and updates and converts them to 'Y', or 'N' accordingly.
Line 7: Consists of a two functions, TRANWRD and COMBL, executed by %SYSFUNC macro function to process a list of variables separated by '#'; replace the '#' with '-'.

See Section 2.9 on Powerful SAS Macro Quoting Functions.

2.5.4 Using the Data Step

By using the Data step to create macro variables, you have great flexibility to combine any combination of the three options below to name and assign values to macro variables. In the general syntax below, target refers to macro variable name, and source refers to macro variable value. These macro variables are created during execution time.

Macro Variable:	**Target Name**		**Source Value**
	Literal String	⟷	Literal String
	Data Step Variables	⟷	Data Step Variables
	Data Step Expression	⟷	Data Step Expression

General Syntax: CALL SYMPUT(TARGET, SOURCE);

```
1. data _null_;
2. set test;
3. if except =. then do;        /* data step condition execution */
4. call symput('newds', new_ds);   /* ('macro variable', dataset variable) */
5. stop;                    /* exit DATA step */
6. end;
7. run;
```

Line 1: The Data _Null_ step may be used to prevent creating a new dataset.
Line 2: Access the TEST dataset.
Line 3: Logical condition needs to be true to execute **Line 4**.
Line 4: The CALL SYMPUT routine is used to create macro variables based on values in dataset. Create NEWDS macro variable based on single observation matching condition. This is important to prevent recreating the macro variable from storing each observation value in the dataset. For multiple records matching **line 3** condition, the value from the last record is saved. In contrast, PROC SQL's INTO: saves the value from the first record. The macro variable value is equal to the dataset variable NEW_DS's value; for example, NEWDS = MYDATA. Note that the CALL SYMPUT routine is an execution time statement to create macro variables.
Line 5: Once the macro variable is created, the STOP statement exits the DATA step. Without the STOP statement, the NEWDS macro variable may be updated with the next record if the condition passes.
Line 6: End of the IF-THEN logic block.
Line 7: End of the DATA step.

```
8. %let rate =.1;
9. data _null_;
10. length dsvar1 $5 dsvar2 4. bill 5.2;
11. dsvar1 = 'mynum';                      /* character value */
12. dsvar2 = 22; hours = 50;               /* numeric values */
```

* In general, there are four types of SYMPUT usage;

```
13. call symput('mvar1','newvalue');              /* literal value */
14. call symput('mvar2', dsvar1);               /* character value */
15. call symput(dsvar1, trim(left(put(dsvar2, 2.))));    /* SAS functions */
```

```
16. call symputx(dsvar1, dsvar2);        /* one char var and one num var */
   /* automatic numeric value conversion to char format before storage */

17. bill = hours*input(symget('rate'), 4.2);
18. run;
```

Line 8: In the general form, the TARGET or the SOURCE may be a literal string (i.e., enclosed in quotes), data step variables, or data step expressions. The LEFT() function realigns numeric values to the left side. The TRIM() function helps to remove trailing blanks on character values. The default format for character values is $W., where W is the width of the variable. The default format for numeric values is BEST12., which is right justified. Note that the macro variable and the dataset variables can have the same name. Note that as regular SAS statements, the CALL SYMPUT can be called with IF THEN conditions.
Line 9: The Data _Null_ step may be used to prevent creating a new dataset.
Line 10: Create three variables — DSVAR1, DSVAR2, and BILL.
Line 11: Assign DSVAR1 to MYNUM value.
Line 12: Assign 22 to DSVAR2 and 50 to HOURS.
Line 13: mvar1 literal is the macro variable name, newvalue is the literal value.
Line 14: Literal name mvar2, dataset variable dsvar1 value of mynum.
Line 15: Dataset variable value mynum as the macro variable name, dataset expression with functions to convert to character using the PUT() function, and remove leading blanks with TRIM() and LEFT() functions. Without these functions, SAS applies default of BEST12. format. Dataset variable dsvar2 has a value of 22. Note that STRIP() is similar to TRIM(LEFT()).
Line 16: Dataset variable value MYNUM as the macro variable name and dataset expression with SYMPUTX() functions to remove leading blanks and convert to character value when assigning 22. Note that SYMPUTX() is a better alternative to **line 15**.
Line 17: Translates to BILL = HOURS*(0.1); this is a similar result as BILL = HOURS*INPUT(&RATE, 4.2); The INPUT() function is needed to convert the character to numeric in the calculation. The SYMGET() function returns macro variable values to the DATA step at execution time, not compile time. This makes the macro variable name more dynamic, since it can be constructed as the program processes each record. This contrasts with directly referencing the fixed macro variable names. With the SYMPUT() function, macro variable names can be constructed and accessed based on data values instead of being hard coded or fixed.

The argument in the SYMGET() function may be a character string (i.e., enclosed in quotes), data step variable, or character expression. In addition, the SYMGET() function is often used to make dynamic changes to the macro variable name and can be used within the same data step as SYMPUT() else; in general, macro variables created by SYMPUT() are only available after the data step that created it. Note that call SYMPUT() and SYMGET() functions must be called within data steps. Also note that PROC SQL supports SYMGET(), but uses INTO: to simulate SYMPUT(). Finally, when applying SYMGET() to save to character variables, the length will be set to 200.

```
19. %let erdst = demog;          /* dataset name */
20. %let ervr = gender;          /* variable name */

21. data _null_;           /* vr_exist created and used in same DATA step */
22. call symputx("vr_exist", varnum(open("&erdst", 'i'), "&ervr"));

23. if input(symget('vr_exist'), 3.0) = 0 then do;
24.   put "*** ABORT program due to variable missing in dataset: &ervr ***";
```

```
25. ABORT;
26. end;                  /* abort if variable does not exist in dataset */
27. run;
```

Line 19: ERDST stores the dataset name.
Line 20: ERVR stores the variable name.
Line 21: DATA _NULL_ since dataset is not created.
Line 22: SYMPUTX call routine calls the VARNUM() function, which returns the number of a variable's position in the dataset or 0 if the variable does not exist in dataset. This is based on the Screen Component Language (SCL) function OPEN(). This value is saved as a character value in the VR_EXIST macro variable. See SCL functions for more info [12].
Line 23: In the same dataset, the VR_EXIST macro variable is retrieved. If the value is 0, meaning the variable does not exist in the dataset, then display user message and abort the program.
Line 24: Display abort message.
Line 25: ABORT program

```
28. data _null_;
29.  set cont;     /* process list of variable names in CLAIMS dataset */

30.  call execute('proc freq data = claims;'    /*execute in middle of DATA step*/
31.  || 'table '              /* table statement for each name value */
32.  || name
33.  || '/list missing; run;');     /* useful common options */
34. run;
            /* translated and executed SAS code per each NAME value */
       /* (  fixed text  )(var_name)(  fixed text  ) */
       /* proc freq data = claims; table claim_var1/list missing; run; */
       /* proc freq data = claims; table claim_var2/list missing; run; */
```

Line 29: Access CONT dataset, which contains the NAME variable that contains the variable names of the CLAIMS dataset. For example, this could be a PROC SQL dictionary dataset of dataset variable names.
Line 30: Creates and executes block of PROC FREQ statements for each value of NAME in each record. CALL EXECUTE routine is useful to create or conditionally execute SAS statements or macro definitions. As a result, in general, there will be SAS statements or macro definitions as arguments.
Line 31: TABLE reserved keyword in PROC FREQ.
Line 32: NAME is the variable in the CONT dataset.
Line 33: LIST and MISSING are other reserved keywords in PROC FREQ.
See also CALL EXECUTE SAS paper [13].

```
35. %macro freqdsn(dsn);
36. proc freq data = &dsn;
37.  tables _all_/list missing;     /* display each variable */
38. run;
39. %mend freqdsn;

40. data _null_;
41.  set datamgt.dbdir;          /* process list of dataset names */
```

```
42. call execute('%freqdsn('          /* execute in middle of DATA step */
43. || name                          /* call to %freqdsn macro */
44. || ')');                         /* for each dataset name */

45. run
          /* translated and executed SAS code per each NAME value */
     /* ( fixed text )(datasn)(    fixed text  ) */
     /* proc freq data = dataset1; table _all_/list missing; run; */
     /* proc freq data = dataset2; table _all_/list missing; run; */
```

Lines 35–39: Defines the FREQDSN macro program.
Line 41: Access DATAMGT.DBDIR dataset, which contains the NAME variable that contains the names of the datasets. For example, this could be a PROC SQL dictionary dataset of dataset names.
Line 42: This is an example of CALL EXECUTE routine used to execute macro definitions. The macro must be quoted. Note that the %FREQDSN macro call is dynamically created and then executed.
Line 43: NAME is the variable in the DBDIR dataset.

2.5.5 Using %Do Loops

```
1. %macro do_loop_ex;
2.   %let start = 1;                  /* local macro variables */
3.   %let stop = 3;
4.   %do age = &start %to &stop;         /* do age = 1 to 3; */
5.     data agetot;                   /*    1st     2nd      3rd      */
6.     set age&age;                     /* set age1; set age2; set age3; */
7.   run;
8.   proc append base = allage data = agetot;   /*append each age1 age2 and age3*/
9.   run;
10. %end;
11. %mend do_loop_ex;
```

Line 1: %DO-%LOOP must be within macro programs.
Line 2: Create START macro variable as 1.
Line 3: Create STOP macro variable as 3.
Line 4: Cycle through **Lines 4** to **8** for each number from 1 to 3. Note that %DO %TO, %DO %WHILE, and %DO %UNTIL behave similar to the DATA Step versions. Do Loops can be used to generate repeated pieces of text.
Line 5: Assign AGETOT dataset.
Line 6: Set AGE&AGE dataset, which translate to AGE1, AGE2, and AGE3 for each do loop iteration.
Line 7: End of DATA Step.
Line 8: Append each new AGETOT dataset to the ALLAGE dataset. With each iteration, AGE1, AGE2, and AGE3 datasets get appended to ALLAGE dataset. Alternative is to set DATA = AGE&AGE directly in PROC APPEND instead of first creating an AGETOT dataset.
Line 9: End of PROC APPEND.
Line 10: End of do loop iteration.

2.5.6 Using PROC SQL and Dictionary Tables

In PROC SQL, the INTO clause is used to create macro variables from summary functions, first row, or multiple rows. Note that the macro variable &SQLOBS is automatically created from a PROC SQL SELECT statement. Also, PROC SQL can be used to access metadata information from DICTIONARY.TABLES. Note that macro variables created by PROC SQL are global unless contained within a macro program.

```
1. proc sql noprint;        /* count the number of records in dataset */
2. select count(old_ds) into :expdcnt from dictexp;
3. quit;
4. %let expdcnt = &expdcnt;    /* useful to remove leading or training blanks */
```

Line 1: Start PROC SQL code.
Line 2: EXPDCNT macro variable resolves to a number, for example 10, based on summary stats COUNT() function. For this example, as an alternative, the &SQLOBS automatic macro variable could be used to get the number of row from the SELECT statement. In addition, the DISTINCT option may also be used to count unique OLD_DS values. Finally, if the COUNT() function was not applied, causing different old_ds values from DICTEXP dataset to be stored in one macro variable EXPDCNT, then only the value from the first record is saved. In contrast to the DATA step, the SYMPUT() function saves the value from the last record.
Line 3: End of PROC SQL.
Line 4: Use the %LET statement to remove any leading or trailing blanks that may exist.

```
5. proc sql noprint;        /* expect multiple values separated by blanks */
6. select old_ds into :expdmvn separated by ' ' from dictexp;
7. quit;
```

Line 5: Start PROC SQL code.
Line 6: EXPDMVN variable resolves to values of OLD_DS separated by blanks, for example, one two three. Without the SEPARATED BY option, only the first value is saved. This contrasts with the CALL SYMPUT, which replaces previous value with the most current value. Note that values can be separated by any character or symbol. In general, a list of values can be processed with the %DO loop or the %SCAN() function. The %SCAN() function is ideal for series of values in one macro variable, since items can be parsed. In general, separate macro variables storing one value each are more flexible than one macro variable storing multiple values.
Line 7: End of PROC SQL.

```
8. proc sql noprint;            /* create three macro variables */
9. select libname, memname, nobs into :lib1, :mem1, :nobs1
10. from dictionary.tables where libname = upcase("&libnm");
11. quit;
```

Line 8: Start PROC SQL code.
Line 9: Multiple macro variables, LIB1, MEM1, and NOBS1, must be separated by commas ','. Note that in this case, only one record is expected.

Line 10: DICTIONARY libname is automatically assigned.
Line 11: End of PROC SQL.

Library, dataset name, and number of records are saved in corresponding variables separated by commas. See SAS website for list of DICTIONARY tables [9].

```
12. proc sql noprint;      /* create three sets of macro variables */
13. select libname, memname, nobs
14.      into :lib1-:lib999, :mem1-:mem999, :nobs1-:nobs999
15. from dictionary.tables;
16. quit;
```

Line 12: Start PROC SQL code.
Line 13: Select LIBNAME, MEMNAME, and NOBS.
Line 14: 1 to 999 multiple macro variables, for each unique value of LIBNAME, MEMNAME, and NOBS. See With a Suffix Indicator using PROC SQL under 2.5 Creating Macro Variables to Store and Replace Text.
Line 15: DICTIONARY libname is automatically assigned.
Line 16: End of PROC SQL.

2.5.7 With a Suffix Indicator Using PROC SQL

In general, there are two methods for creating macro variables with suffix indicators — proc sql and data step. Typically, these macro variables will be referenced as &&DSN&I within a %DO I = 1 %to &N; … &&LAB&I … %END loop.

2.5.7.1 Using PROC SQL

```
1. proc sql noprint;         /* predetermine number of macro variables */
2. select label into :lab1 - :lab&expdcnt /* :lab999 for unknown # of values */
3. from dictexp order by old_ds;
4. quit;
```

Line 1: Start PROC SQL code.
Line 2: Create macro variables lab1 to LAB&EXPDCNT. Make sure to %LEFT %TRIM(&EXPDCNT)) in advance. Each macro variable has a suffix indicator that needs to be resolved using double && (ex. &&LAB&I) if the suffix indicator is not explicitly specified. Each macro variable value is stored from each row. Note that for multiple rows, only the first row is saved when creating only one macro variable. Note that the auto macro variable &SQLOBS can be used, with the %DO loop to cycle through the list of macro variables. This process is similar to creating an array and storing values in each array element.
Line 3: ORDER BY allows for more control and ordered or macro values.
Line 4: End of PROC SQL.

2.5.7.2 Using the data step

```
1. data _null_;            /* create macro variables */
2.  length j $2.;
3.  set sashelp.class;
4.  i + 1;
```

```
5. j = left(put(i, best12.));
6. put j = ;
7. call symput('n', j);           /* final n = total number of records */
8. call symput('name'||j, trim(left(name)));   /* name1 = John name2 = Mary */
9. run;                           /* Macro variable name = Value */

10. %macro displayv;              /* reference list of macro variables */
11. %put n = &n;
12. %do i = 1 %to &n;             /* loop through total number of variables */
13. %put name&i = &&name&i;      /* display name1 = John name2 = Mary etc. */
14. %end;                  /* && is needed for double resolution */
15. %mend displayv;
16. %displayv;
```

Line 1: The Data _Null_ step may be used to prevent creating a new dataset.
Line 2: Create variable J.
Line 3: Set SASHELP.CLASS dataset.
Line 4: Counter variable I to store sequence number from 1.
Line 5: Format and left align counter variable using BEST12. format.
Line 7: Create macro variable N to store the total number of entries. Note that this value gets updated with each DATA step iteration.
Line 8: Create macro variables NAME1 to NAMEXX for each NAME value in the dataset CLASS. The J character variable assigns a sequence number to each dataset name. The order stored is based on the dataset order. Note that in version 9, SYMPUTX() also trims leading and trailing blanks.
Lines 10–15: Generally, a %DO-%LOOP within a macro program is applied to access/display the values of the list of macro variables.

2.5.8 Confirming and Debugging Macro Variable Values

Once macro definitions are accessed or loaded, in general, there are five types of macro programming errors that need to be addressed:

2.5.8.1 Compile time errors

1. Correct macro syntax requires confirming macro programming rules are applied.
2. Open code, statements outside of macro definitions, issues can be prevented by confirming all open code macro statements are valid.

2.5.8.2 Execution time errors

1. Unresolved macro variables can be prevented by confirming the scope (global, local) and assignment of macro variables.
2. Correct resolved macro syntax requires consistency between data and variable type.
3. Correct macro programming logic requires confirming start/end of %DO LOOPS, %IF/%THEN conditions, operators, and comparisons for numeric, character, and date values.

2.5.8.3 Resolved macro code

Note that macro-generated errors generally do not have line numbers, as compared to other SAS ERRORs or WARNINGS. See the SAS paper on debugging macro programs [14].

The %PUT statement can be used as open code to display macro values in the log file. Quotes are not needed. Along with constants, %PUT can also call macro programs that dynamically list macro variables (see Table 2.1).

%put Value of '&newds' macro variable is &newds., and firststy is &&firstv&numfv..;.

Value of &newds macro variable is mydata, and firststy is myage.

Note that single quote macro variables, such as '&CITY,' are not resolved, and double quotes, such as "&CITY," are required when using macro variables in titles and footnotes. Dynamically create code incrementing starting to ending macro variables.

```
1. %macro macprocess;          /* reference list of macro variables */
2. %do i = 1 %to &onam_cnt;
3.  %put i = &i nam&i = &&nam&i ;  /* && is needed for double resolution */
4. %end;
5. %mend macprocess;
   i = 1 nam1 = myname
   i = 2 nam2 = newname
   ...
```

The %DO %END Loop can be used to display multiple macro values. When using the %DO %END Loop, the macro reference is typically &&NAM&I. See Table 2.1 Indirect References to Macro Variables.

Additional tools to help debug macro variables and macro programs include these options: MPRINT (log shows the code actually created by macro), MLOGIC (log shows flow of macro conditional execution), MERROR and SERROR to display warnings, and SYMBOLGEN (log shows macro variable value). MPRINT and MFILE options can be used to create a macro-free version of the code. In addition, the automatic macro variables _ALL_, _LOCAL_, _GLOBAL_, _USER_, or _AUTOMATIC_ can be displayed. See SAS website for a complete list of automatic macro variables [9]. See SAS paper on unraveling macros [15].

```
1. filename mprint 'C:\macros\mymac_unravel.sas'; /* new unmacro version code*/
2. options mprint mfile;              /* print to file macro options */

3. %macro mymac(libn, dsn =);     /* load one or more nested macro programs */
4. proc print data = &libn..&dsn;run;
5. %mend mymac;

6. %macro unravel();             /* wrap the macro call in another macro */
7. %mymac(mydata, dsn = demog);
8. %mend unravel;
9. %unravel;                        /* call macro */
```

TABLE 2.1 Indirect references to macro variables

```
* Indirect reference to one macro variable – use &&&;
```

Steps

```
%let dog = hound; /* macro value of final resolution */
%let pet = dog; /* name of macro variable */
%put 'My &pet = ' &pet 'is a &&&pet = ' &&&pet;
```

1

```
My &pet = dog is a &&&pet = &dog;/* &pet resolves to
    dog */
                                  /* && resolves to & and
                                     &pet resolves to
                                     hound */
```

2

```
My &pet = dog is a &&&pet = hound; /* rescan &dog to
   hound */
```

```
* Indirect Reference to list of ROOT Macro Variables - use
  &&XX&;
```

Steps

```
%let dsn = animals;/* not required since not resolved */
%let n = 2;         /* index value for list of macro
                       variables */
%let dsn2 = cats;   /* ROOT+<INDEX_NUMBER> macro value */
%put '** &dsn&n = ' &dsn&n ' ** &&dsn&n = ' &&dsn&n;
```

1

```
** &dsn&n = animals2 ** &&dsn&n = &dsn2; /* &dsn&n
   resolves to animal2 */
                                  /* && resolves to & and
                                     &n resolves to 2 */
```

2

```
** &dsn&n = animals2 ** &&dsn&n = cats; /* rescan &dsn2
   to cats */
```

```
* Indirect Reference to list of ANY Macro Variables - use
  &&&XX&;
```

Example

```
%LET Engine1 = Make Model EngineSize;
%LET Engine2 = Make Model Cylinders;
%LET MPG1 = Make Model MPG_City;
%LET MPG2 = Make Model MPG_Highway;

%MACRO triple(var);
  %DO i = 1%TO 2;
    Title "Cars: &var &i";/* &VAR and &I are used
       separately */
    PROC PRINT DATA = sashelp.cars;
     VAR &&&var&i.;/*Resolve &Engine1, &Engine2,
        &MPG1,%MPG2*/
    RUN;
  %END;
%MEND triple;
%triple(Engine);
%triple(MPG);
```

1

```
&&&var&i. /* && resolves to &, &var resolves to Engine,
   &i resolves to 1 */
```

2

```
&Engine1 /* rescan &Engine1 to Make Model EngineSize */
```

Line 1: Use FILENAME to point to new unmacro version SAS program name.
Line 2: Options MPRINT and MFILE will turn on writing any SAS code executed to FILENAME file.
Lines 3–5: MYMAC macro program is loaded or accessed from SASAUTOS option.
Lines 6–8: Wrap the MYMAC macro program call within another macro program, UNRAVEL.
Line 9: Executing the %UNRAVEL macro will save the SAS code executed from the %MYMAC macro program to the MYMAC_UNRAVEL.SAS program.

2.5.9 Correcting Common Macro Problems

Assure code runs fine if submitted as open code. If not, then error messages mean that tokenization needs to be corrected due to syntax error. Possible solutions include missing semicolon, parenthesis, or quotation mark, or unclosed comments. Other options may be missing parameter when called or incorrect logic in the code.

2.6 REFERENCING MACRO VARIABLES TO SUBSTITUTE TEXT

When using macro variables, all references to the macro variables will be replaced by the substitute value before the code is compiled.

2.6.1 In SAS Statements

In general, there are five ways for referencing macro variables, including in conditions, formats, assignments, lists ,or conditional statements.

1. As values in conditions

   where old_ds = “&newds”;

   Resolved: where old_ds = “mydata”;

   For full replacement in any procedure. Double quotes remain and are used, since performing a character comparison. Note that single quote macro variables will not resolve.

   if “&studyn” = study then do; ...; end;

   Resolved: if “2001123” = study then do; …; end;

Macro could be number value and compared to a character variable type. Actual value of study variable should be numeric. Notice no “” around the study variable.

2. As formats

```
%let cyc = 1;
proc format;
 value cyc&cyc.f
     &cyc = “Cycle &cyc”;
run;
```

```
Resolved:
proc format;
 value cyc1f 1 = “Cycle 1”;
run;
```

Double quotes remain.

3. As variable assignment (num or char)

```
age = &myage;
Resolved: age = 50;
```

Assign value from macro variable to age variable. Age is a numieric variable.

```
study = “&mystudy”;
Resolved:  study = ‘Protocol A’;
```

```
study = “&studyn”;
Resolved:  study = ‘2001124’;
```

Assign value from macro variable, character or numeric value to study, which is a character variable. Character variables are assigned with double quotes which get resolved to single quotes.

4. As list of variables

```
var %varlst;
Resolved: var sales2000 sales2001 sales2002;
```

Macro could loop through a set of values to create a list of related variables.

5. As conditional statements

```
if beg&cycle.dt ^ =.;
Resolved: If beg1dt ^ =.;
```

The dot in the .DT defines when the macro variable name stops, and the DT is extra text added to the macro variable’s value.

```
if beg %eval(&cyc + 1)dt ^ =. then do;
```

Resolved:
If &cyc = 1;
If beg2dt ^ =. then do;

The %EVAL() function preprocesses the expression by adding the value of &CYC to 1 before resolving the BEG2DT variable in the condition. %EVAL() is used for integer math.

2.6.2 As Two or More Macro Variables in a Sequence

1. set c&lbname..&dsn;

Resolved: set cmylib.mydata;

Line 1: Two .. are needed between the macro variables &LBNAME and &DSN to resolve the two macro variables independently and to include one . to resolve to the correct syntax. The '.' special character is treated as part of the macro variable and does not appear when macro is resolved.
Line 2a: libname cdata2 'C:\mydata';
Line 2b: %let n = 2;
Line 2c: %let lbname2 = data2;
Line 2d: set c&&lbname&n...mydata;

First pass resolve: set c&lbname2..mydata;
Second pass resolve: set cdata2.mydata;

Line 2d: Three '...' are required because two passes are needed to resolve c&&lbname&n. The first pass compresses the three periods to two, and the second pass compresses the two periods to one. The objective is to resolve to the predefined libname.

2.6.3 As Two or More Ampersands (&&) or Indirect Macro References

1. %let dog = hound;
2. %let pet = dog;/* macro variable value is name of **line 1** macro variable */

3. %put %nrstr(My &pet =) &pet %nrstr(is a &&&pet =) &&&pet;
   /* Resolves to My &pet = dog is a &&&pet = hound */

4. %put My &pet is a &&&pet; /* indirect macro reference */

First pass: My &pet = dog is a &&&pet = &dog; /* &pet resolves to dog */
/* && resolves to & and &pet resolves to dog */
Second pass: My &pet = dog is a &&&pet = hound; /* rescan &dog to hound */

Line 3: %NRSTR() masks the &PET and &&&PET tokens to display as is equal to the resolved values.
Line 4: &pet resolves to dog. First resolution of &&pet is &dog, and then the second to hound. Double ampersands are used to resolve twice before execution. Note that this

process is repeated until all '&' values are resolved. In general, limit the ampersands (&)s to prevent confusion.

```
4. %let dsn = animals;       /* root macro variable name */
5. %let n = 2;          /* suffix number */
6. %let dsn2 = cats;         /* root with suffix number as new macro variable */

7. %put '** &dsn&n = ' &dsn&n ' ** &&dsn&n = ' &&dsn&n;

First pass:    ** &dsn&n = animals2   ** &&dsn&n = &dsn2; /* second builds
    macro var */
Second pass: ** &dsn&n = animals2   ** &&dsn&n = cats;     /* rescan to new
    macro var */
```

Each individual (&) resolves immediately, while the two (&&) resolve to one (&). This process is repeated until all (&)s are resolved. This technique is often used with %do loops with a root macro variable name and suffix number. In general, macro variable lists with total number of items need to be created in advance before resolution. The first macro reference, '&dsn&n', concatenates the root and suffix number to resolve to ANIMALS2. The second macro reference with the extra '&', '&&dsn&n', resolves to the new root with suffix number macro variable. As an alternative to %DO loop processing with suffix numbers, you can use %SCAN() to parse items from one macro variable containing a list of values. The single quotes in the %PUT statement keep macro variable names as literals.

See Table 2.1 Indirect References to Macro Variables.

2.6.4 With Macro Functions

Macro functions share the same basic syntax and purpose as their corresponding DATA step function, except that they are applied to macro variables to preprocess the value before resolution. Generally, there are two types of macro functions — rescan and no rescan. No rescan macro functions prevent SAS from repeating the process of translating special macro characters, such as '&' and '%', and treating them as literals. By not rescanning selected SAS code, SAS compiles the code and only executes it when the condition applied is true.

See SAS website for list of SAS supplied auto call macro functions [9]. See Using Macro Functions under 2.5 Creating Macro Variables to Store and Replace Text for more information. See SAS paper on creating lists in macro variables [16]. See SAS paper on looping through values in macro variables [17]. See SAS paper on accessing file path names [18].

2.6.4.1 For general manipulating character strings

```
1. %let name = John Doe;
2. %let upname = %upcase(&name);
3. %let upname = JOHN DOE;
/*converts letters from lowercase to uppercase */
```

```
4. %let num = %length(&name);
5. %if %length(&name) > 0 %then %do;
6. %end;
7. %let num = 8;
/* returns the length of a character string */

%let fstname = %substr(&name, 1, 4);
%let fstname = John;
/* returns a substring of a character string */

%let lstname = %scan(&name, 2);
%let lstname = Doe;
/* extracts a word from a character string */

%let mlnam = acc.dth.dis;
/*macro variable containing list of tokens*/

%let nval = 1;
/* Create counter macro variable */

%let clab = ;
/* Create token macro variable */

data supupdate;

%do %while(%length(%scan(&mlnam, &nval, %str('.')))> 0);
/* Do Loop through each token in the list until last token is read */

%let clab = %scan(&mlnam, &nval, %str('.'));
/* Assign token value based on order in list */

%put clab = &clab;
/* Display token value */

liv&clab = 1;
/* Create and assign value = 1 */

&clab.rt = 100*liv&clab;
/* Reference token value in statement */

%let nval = %eval(&nval + 1);
/* Increment counter by 1 */

%end;
run;

livacc = 1; accrt = 100*livacc;
livdth = 1; dthrt = 100*livdth;
livdis = 1; disrt = 100*livdis;
/*%Do% While Loop processes the MLNAM macro variable for each value. The SAS
   statement within the loop is repeated three items — once for each item in the
   MLNAM macro. */
```

```
%let srchname = %index(&name, John);
%let srchname = 1;
/* returns the first occurrence position number of a character string within another
   string */

%let street = 123 Main St. ;
%let street = %cmpres(&street);
%let street = 123 Main St.;
/* multiple blanks are compressed to a single blank and leading and trailing blanks
   are removed. */
```

2.6.4.2 Within a character variable IF condition

```
1. proc sql;
2.  select count(distinct sex) into: sexn from sashelp.class;
3.  select distinct sex into: sexlst separated by ' ' from sashelp.class;
4. quit;

5. data nonmisssex misssex;
6.  set sashelp.class;
7.  if sex in (%do sext = 1 %to &sexn; "%scan(&sexlst, &sext,' ') "%end;)
8.  then output nonmisssex;
9.  else output misssex;
10. run;
```

Line 2: Counts the number of unique SEX values in SEXN macro variable.
Line 3: Saves a list of unique SEX values in SEXLST macro variable.
Line 7–8: Creates a new macro variable SEXT within the %DO %LOOP. Processes the list in SEXLST macro variable one value at a time to resolve to — IF SEX IN ("MALE" "FEMALE") THEN OUTPUT NONMISSSEX.
Line 9: Saves all missing SEX value records in MISSSEX dataset.

2.6.4.3 For manipulating character strings using quoting functions

```
1. %let prnt = %str(proc print; run;);

2. %put %nrstr(proc print data = &study..demog; run;);
   /* Resolved to display proc print data = &study..demog; run; */

3. %let title1 = %bquote(Summary of N (#) of Adverse Events);

4. %let title2 = %nrbquote(Summary of N (%) of Adverse Events);
5a. %let study = 443;
5b. %put %nrbquote(proc print data = &study..demog; run;);
   /* Resolved to display proc print data = 443.demog; run; */

6. %let titl = %nrbquote(^S = {font = ("arial", 11 pt)}Table 14-2. Demographics and
     Baseline Characteristics);
```

```
7. titlel "&titl";

                (     1st token     )              (   2nd token   ) (3rd token)
8. &_titl = 'ABC Pharmaceuticals, Inc. ' || "%nrbquote(")" || ' j = r '
|| "%nrbquote(")" || ' Page ~{pageof}';
  (4th token )        ( 5th token )

                (     1st token     )     (   2nd token   )  (3rd token)
9. &_tit2 = "ABC &_cpmstudyt" || "%nrbquote(")" || ' j = r ' ||
"%nrbquote(")" || ' Report';
( 4th token )      (5th token)

10. %let lcode = %quote(123 O%'Connor Street);         /* similar to %STR() */
11. %let lcode2 = %nrbquote(123 O%'Connor &Street);  /* similar to %NRSTR() */

12. title "Current Date %sysfunc(left(%qsysfunc(today(), worddate18.)))";
```

Lines 5a–b: Resolves &study, but still does not execute since often used with %PUT statement.
Line 8: 5 Tokens resolve to TITLE1 "ABC Pharmaceuticals, Inc. " J = R " Page ~{pageof}";
Line 9: 5 Tokens resolve to TITLE2 "ABC 416858-CS01 " J = R " Report";

In general, the corresponding no rescan version functions behave similarly except that (&) and (%) are also ignored. See Using Macro Functions under 2.5 Creating Macro Variables to Store and Replace Text. See Section 2.9 Powerful SAS Macro Quoting Functions. See SAS macro quoting SAS paper [11].

2.6.4.4 For arithmetic and logical operations

```
1. %let numer = 4;
2. %let denom = 2;
3. %let result = %eval(&numer/&denom);     /* %let result = 2; */

/* returns the arithmetic value of 4 divided by 2. Noninteger source values (&numer,
&demon) and results are truncated to integers. Use %SYSEVALF() to calculate
decimals. Without the %EVAL(), result would be equal to 4/2. Any operator can be
specified. Use %SYSEVALF() function for noninteger expression*/

4. %let test = %eval (5 < 10);
5. %let test = 1;

/* returns 1 (true) instead of 0 (false) based on the logic operation. Any noninteger in
expression results in character comparison and not numeric. */
```

2.6.4.5 Outside of data steps

```
footnote "On %sysfunc(today(), mmddyy10.) ";
```

%SYSFUNC() allows for functions outside of DATA steps.

2.6.4.6 *For accessing system environmental variables*

```
1. %macro getpathname;
2. %sysget(SAS_EXECFILEPATH)
3. %mend getpathname;
4. %put %getpathname;          /* returns c:\mymacros\mymac.sas */

5. %macro getfilename;
6. %qsubstr(%sysget(SAS_EXECFILENAME),1, %length(%sysget(SAS_
   EXECFILENAME)) -4)
7. %mend getfilename;
8. %put %getfilename;          /* returns mymac */
```

Line 1: Need to enclose in a macro program.
Line 2: %SYSGET() accesses the full path name. This is similar to SYMGET(). The system environmental variable name may be dependent on the operating system.
Line 5: Need to enclose in a macro definition.
Line 6: %SYSGET() accesses the file name. See SAS paper [18]. The system environmental variable name may be dependent on the operating system.

2.6.5 To Conditionally Create SAS Code or Statements Using a List of Values as an Option

In general, many options exist to conditionally pass SAS code to the compiler, including constant value, macro variables, nonmissing value, and indexed strings. MLOGIC option is useful for debugging purpose. Note that %GOTO statements can also be used to transfer control to labeled portions of the macro. See SAS paper on %GOTO statements [19].

2.6.5.1 *Creating a dataset based on a macro variable equal to 0 or 1 condition*

```
%macro xxx;
1. %if &expdcnt = 0 or &expdcnt = 1 %then %do;
2.  data test;
3.  set dsname;
4.  ...;
5. run;
6. %end;

7. %if (&expdcnt ne) %then %do;    /* confirming non-missing macro value */
%mend xxx;
```

Line 1: Note that there are no quotes around 0 or 1. Number or character values can be compared. Note that the IN operator in %IF %THEN statements are not valid. Note that the expression %IF 1.0 = 1.0 is true, but the expression %IF 1 = 1.0 is false. Note that the IN operator does not work with macro programming.
Lines 2–5: Any block of SAS code to execute based on **line 1** being true.
Line 6: end of %IF%THEN block.å
Line 7: As an alternative, the %IF %THEN statement can check for nonmissing macro values.

Macro variables enable conditional execution of data or PROC steps, SAS statements, or parts of SAS statements. Make code compile or not based on conditional %IF %THEN statements. The result of the %IF %THEN statement determines if the SAS compiler will see the code, as compared to the IF THEN block of code will be executed when using the IF THEN statements. SAS compiles all code when using the IF THEN block of code. In general, the %IF-%THEN statements should be based on user-defined or system-based values, since DATA step variable values are only available during execution time that uses IF-THEN statements.

Using the macro variable &EXPDCNT resolves to a number and compares to 0. Equality and inequality comparisons take place between the resolved value of an expression on both left and right sides. The value returned by the comparison is 1 (true) or 0 (false). Notice no quotes around macro variable or 0. Almost any code or symbol, such as ' ' ;or. ;, can be created. See SAS paper on comparing IF and %IF [20].

```
7. %if %length(&expdcnt) ne 0 %then...;
```

As an alternative, %LENGTH() function may be used to confirm nonmissing macro value.

```
7. %if %eval(&expdcnt) = 2 %then...;
```

As an alternative, %EVAL() function may be used as calculation operators.

```
1. %macro filter(expdcnt)/minoperator;
2. %if &expdcnt in 0 1 %then %do;
3. ...
4. %mend filter;
```

As an alternative, with the MINOPERATOR option and MINDELIMITER = space default option, it is possible to apply the IN operator in macro programming. This technique also works for character values. Note that quotes should not be included. See SAS paper for example [21].

2.6.5.2 Printing a dataset based on a macro variable equal to 0 condition

```
%macro xxx;
1. %let expdcnt = 0;
2. %if &expdcnt = 0 %then proc print data = test;;      /* SAS statement */
3. %if &expdcnt = 0 %then %str(proc print data = test; var name sex; run;);
%mend xxx;
```

Another way to execute code based on a macro variable is to specify the SAS statement after the %THEN. Note that only one SAS statement with two ';' semicolons are required. Multiple SAS statements can be included with the %STR() function. Any macro expression can be specified.

2.6.5.3 Assigning a new macro variable based on another macro variable condition

```
%macro xxx;
1. %if &expdcnt = 0 %then %let prtchk = YES;        /* new macro variable */
2. %else %let prtchk = NO;
%mend xxx;
```

%LET can be conditional executed only with the %IF-%THEN statement, or else the second %LET statement will overwrite the first %LET statement, since this is a compile time statement. See SAS paper on %IF and IF differences [20].

2.6.5.4 Creating a macro variable based on a macro variable matching condition

```
%macro xxx;
1. data _null_;
2. set __dmp;
3. %do i = 1 %to &onam_cnt;     /* Loop through list of macro variables */
4.  %if &&onam&i = &&nam&i and    /* Preprocess to decide when to compile */
5.  %upcase(&&onam&i) ^ = PATIENT %then %do;
6.     if upcase(name) = "&&onam&i" then do; /*Always compiled, data value*/
7.        call symput("otyp&i", type); /* Create macro variable */
8.     end;
9.  %end;
10. %end;
11. run;
%mend xxx;
```

Lines 3–10: %DO %END Loop is used to create SAS code. %DO %END Loops work similar to their corresponding DATA step statements. Three sets of macro variables are processed at the same time. %IF statement compares two macro variables to each other and one macro variable with a constant. Notice no quotes around macro variables or the constant. Resolves to a simple Boolean expression.

Double && ampresand signs are used and resolved as follows:

&ONAM1 (resolve suffix indicator, first & is used)
OLDDS (value of &ONAM1, second & is used)

Actual SAS code block is created for each macro variable to compare each of the dataset variable names with each macro variable value. Once a match is found, a new macro variable &OTYP&I is created using the CALL SYMPUT() function. This is a dynamic way to create macro variables based on values in dataset. Condition must select only one observation in the dataset.

Also %IF ("&STATUS" NE "END") %THEN %DO;

See Table 2.1 Indirect References to Macro Variables.

2.6.5.5 *Dynamically executing a macro based on a macro greater than missing condition*

```
%macro xxx;
1. data _null_;
2. set __dmp;
3.
4. %do i = 1 %to &onam_cnt;
5.  %if &&spec&i > ' '%then %do;
6.   %&&spec&i.(&&nam&i, &&onam&i);
7.  %end;
8. %end;
9. run;
%mend xxx;
```

%if %then statement compares macro variable to missing value. If macro variable has a nonmissing value, then the macro resolved by &&SPEC&I will be executed. The '.' is needed after the I and before the '(' to specify the end of the macro variable name I. Without the '.', SAS may get confused, since a macro variable name cannot contain special characters.

Double && ampersand signs are used and resolved as follows:

1. &SPEC1 (resolve suffix indicator, first & is used)
2. U_CONVCD (value of &SPEC1, second & is used)

The macro variables passed as parameters are also using double && that resolve as follows:

1. &NAM1, &ONAM1 (resolve suffix indicator, first & is used)
2. NEWNAM, OLDNAM (value of &NAM1 and &ONAM1, second & is used)

See Table 2.1 Indirect References to Macro Variables.

2.6.5.6 *Executing a macro based on a macro function condition*

```
%macro xxx;
1. data _null_;
2.  set —dmp;

3.  %do i = 1 %to &onam_cnt;
4.  %if %index(%substr(&&spec&i, %length(&&spec&i) - 2), YN) > 1 %then %do;
5.         %u_convyn(&&nam&i, &&onam&i);
6.  %end;
7.  %end;
8. run;
%mend xxx;
```

Line 4: Any valid expression can be used in the condition. %INDEX(), %SUBSTR(), and %LENGTH() functions work the same way as their corresponding functions. Ex. the length function returns the length of the character value. Ex. length of SMOKEYN is 7. For null values, %LENGTH returns 0.

See Table 2.1 Indirect References to Macro Variables.

2.6.5.7 *Executing a macro within a do loop and scan function*

```
%macro xxx;
1. %do kk = 1 to %wrdcnt(&varlist, '#');
2. %mcon(patid = subjid, convar = %scan(%quote(&varlist), %eval(&kk), '#'));
3. %end;
%mend xxx;
```

2.6.5.8 *Constructing a keep statement based on a do loop*

```
%macro xxx;
1. data final;
2.  set final;
3.  keep
4.  %do i = 1 %to &onam_cnt;
5.    %if %index(&&nam&i, _) < 1 %then %do;
6.      &&nam&i %substr(&&nam&i, 1, (%length(&&nam&i) - 2))di
7.    %end;
8.  %end;
9. ;
10. run;
%mend xxx;
```

Line 3: constructing a KEEP statement.
Line 4: do loop processes from 1 to total number of macro variables. Within the do loop, macro processor resolves all direct and indirect macro variables until loop is completed before compiling and executing code.
Line 5: macro %IF %THEN determines if '_' is part of the &&NAM&I macro variable name. If not, then execute **line 7**.
Line 6: If the macro variable &&NAM&I does not have the _ in it's value, then it will be listed in the keep statement. In addition, a modified version of the macro value will also be listed. i.e., the last two suffix character values are replaced with the di suffix. Each macro variable that meets the %INDEX() function condition will be processed.
Line 7: end of %IF %THEN block.
Line 8: end of %DO loop.

See Table 2.1 Indirect References to Macro Variables.

2.6.6 To Conditionally Create SAS Code or Statements through a List of Variables in One Macro Variable

```
%macro xxx;
1. %if &expdcnt = Y or &expdcnt = Yes %then %do;
2.  data test;
```

```
3.  set dsname;
4.  ...;
5.  run;
6. %end;
%mend xxx;
```

Lines 1–6: Similar to the corresponding numeric values, character values can also be applied in %IF-%THEN statements. Note that quotes should not be applied. Note that MINOPERATOR option and the MINDELIMITER = space default option can also be applied to simulate the IN operator.

2.6.7 To Write SAS Code That Creates SAS Macro Programs

Three elements of SAS Code generators:

2.6.7.1 Access metadata files

a. Dynamically create and store macro variables containing one or list of values
b. Selectively applying macro variables as individual values or as a unit list of values

```
1. %let state = %nrstr(&CA);
2. %put I live in %superq(state).;                /* I live in &CA. */

3. %let state = &CA;
4. %let ca = my state;
5. %put I live in &state.;                 /* I live in my state. */

6. filename newmac 'C:\mymacros\newmac.sas';        /* write to newmac.sas */
7. options linesize = 256;
8. data _null_;
9.  file newmac;
10. put '%macro mymac(libn, dsn =);';           /* macro syntax within put */
11. put ' proc print data = &libn..&dsn; run;';  /* simple fixed sas code */
12. put '%mend mymac;';
13. run;

Resolved newmac.sas file -
%macro mymac(libn, dsn =);
proc print data = &libn..&dsn; run;
%mend mymac;

14. filename newmac 'C:\mymacros\newmac.sas';       /* write to newmac.sas */
15. options linesize = 256;
```

```
16. %let libv = sashelp;     /* macro variables to be used in data null step */
17. %let varn = 2;           /* total number of variables in var statement */
18. %let var1 = sex;         /* first variable */
19. %let var2 = weight;      /* second variable */
20. data _null_;
21. file newmac;
22. put '%macro mymac(libn, dsn =);';          /* macro syntax within put */
23. put '%if &libn = ' "&libv " '%then %do;';     /* conditional execution */
24. put ' title ** SASHELP Libname ** "&dsn" ;';
25. put ' proc print data = &libn..&dsn;';
26. put '   var ';
27. %do i = 1 %to &varn;
28. put "&&var&i";                 /* first translates to var var1 var2; */
29. %end;                  /* second translates to var sex weight; */
30. put ';';
31. put '   run;';
32. put '%end;';
33. put '%mend mymac;';
34. run;
```

```
Resolved newmac.sas file -
%macro mymac(libn, dsn =);
%if &libn = sashelp%then%do;
  title ** SASHELP Libname ** "&dsn" ;
  proc print data = &libn..&dsn;
  var
sex
weight
;
  run;
  %end;
%mend mymac;
```

Lines 1–2: With the NRSTR() and SUPERQ() macro quoting functions, the '&' special character is preserved without resolving the macro variable. Note that the SUPERQ() function does not contain the '&' symbol. Note also that without the NRSTR() function, SAS will issue a warning message, since it will first try to resolve &CA. This unique feature allows you to store macro variables or macro programs within macro variables for display process and without requiring resolution.
Lines 3–4: Without the NRSTR() and SUPERQ() macro functions, SAS always resolves macro variables.
Line 6: Assign FILENAME statement to SAS program name.
Line 9: Writes **Lines 10** to **12** as a simple fixed SAS code example to NEWMAC.SAS file.
Lines 10–12: Example of simple fixed SAS code.
Line 14: Assign FILENAME statement to SAS program name.
Lines 16–19: Assign values to four macro variables — LIBV, VARN, VAR1, and VAR2.

Line 20: The Data _Null_ step may be used to prevent creating a new dataset.
Line 21: Writes **lines 22** to **33** as conditional execution to NEWMAC.SAS file.
Lines 22–33: Series of PUT statements with SAS code in quotes including ';'.
Line 23: The &LIBV macro resolves to SASHELP in the statement — &IF &LIBN = SASHELP %THEN %DO;.
Line 24: Notice that the TITLE statement is created and that double quotes "" are before and after &dsn macro variable.
Lines 27–29: DO LOOP process to iterate two times to display two macro variables in the statement — VAR SEX WEIGHT;.

2.7 THE ART OF SAS MACRO TESTING

2.7.1 General Process

1. Create macro utility program as instructed.
2. For best practices, use PROCOPTSAVE and PROCOPTLOAD to save and restore user SAS systems.
3. Validate macro utility program; fill out QC excel file.
4. Check program header, section comments, SAS log, etc.
5. Is each parameter documented?
6. In general, are keyword parameters used instead of positional parameters?

2.7.2 Suggested Guidelines for Validating Macro Utility Program

1. Create test dataset with test patients based on macro function and code review.
2. Include good and bad data values in the test dataset.
3. Include duplicate records for selected cases in test dataset.
4. If possible, also use CRTs with real data when testing.
5. In general, you should not have to independently write a parallel macro program to compare results with the production macro.
6. If possible, offer suggestions for improving or resolving identified issues.

2.7.3 Can You Call the Macro without Writing to the Production Library?

1. If so, then this is the best/easiest approach: Call macro for each test case.
2. If not, then copy/paste macro into test macro; do not write to library.

2.7.4 Matrix of Test Cases—Excel File of All Possible Combinations

1. First row — name of each macro parameter.
2. Second row — brief description of each macro parameter.
3. Third row — list of default values, if any, for each macro parameter.
4. Fourth row — list of valid values for each macro parameter, one row for each valid value.
5. Fifth row — list of invalid values for each macro parameter, one row for each invalid value.
6. All combinations of unique valid (sixth row) and unique invalid (seventh row) macro parameter values.

2.7.5 Consider These Factors When Developing Test Cases

1. Check providing/not providing required/optional parameters.
2. Check for character/numeric variable type.
3. Check for User Messages for ERRORs, WARNINGs, and Critical Notes.
4. Individually test for each valid option for each parameter. For example, two options for two parameters each result in four test cases.
5. Check for invalid dataset/variable/value in parameter.
6. Check for subset resulting in zero observations.
7. Check for all, some, and no missing values.
8. Check for all, some, and no negative values.
9. Check unit and integrated parameters.
10. Check single/multiple variables/values specified in a parameter.
11. Check format parameters (example: landscape, column width, etc.), if applicable.
12. Check all limited possible combinations of macro parameters.
13. For each test case, assure all observed values match all expected values.
14. Track changes in record counts of intermediate datasets to assure no loss of data.
15. Check any combination of the above factors as needed.

2.7.6 Final Steps

1. Verifying cleanup and reducing possible interference with other parts of the calling program.
2. Delete any datasets it creates.
3. Restore the SAS autos search path or other important options if changed by macro.
4. Be aware of the effect of any global macro variables defined.
5. Only create datasets starting with _ (or have another way to avoid name conflicts with existing data).

2.8 SECTION SUMMARY

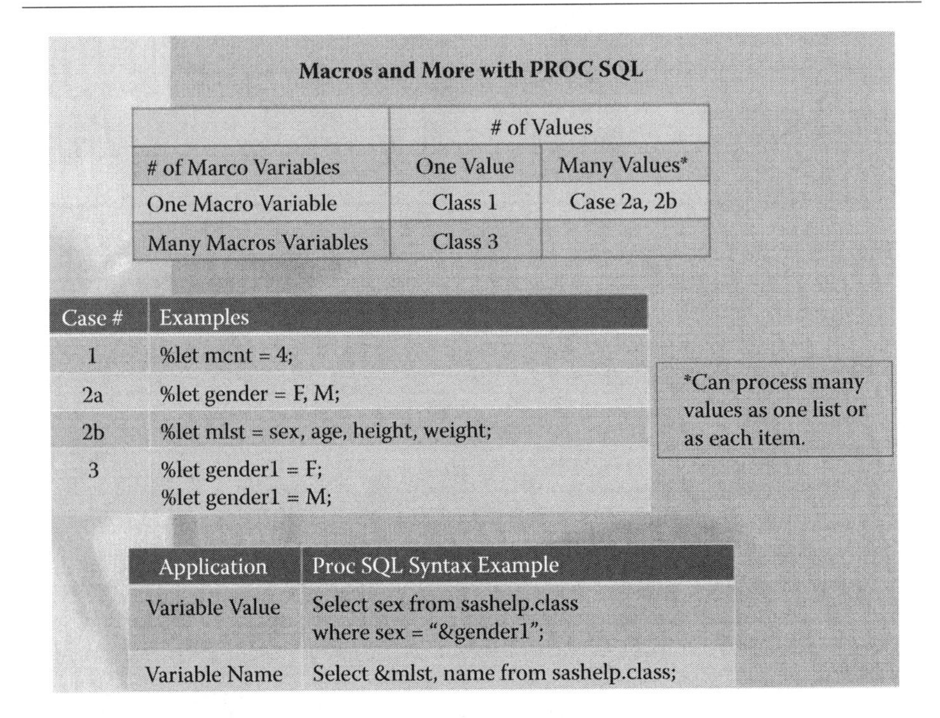

See the following tables for quick reference to SAS Macro Programming.

Table 2.2 Type of macro token values
Table 2.3 Useful automatic SAS macro variables
Table 2.4 Useful debugging SAS macro options
Table 2.5 Selected open code SAS macro statements

TABLE 2.2 Type of macro token values (all characters)

MACRO VARIABLE NAME	*MACRO TOKEN VALUE (MOST ANY VALUE)*
dataset names	class, shoes, demog, mylib.demog
character variables	name, gender
numeric variables	age, weight
date variables	dob, aestdt
data values — chr, num, date	SC, " hello ",1, 2001123, '01JUN2010'D
token values or macro variables	mylib.mydata, city center gender, &dsn (obs = 10)
special characters (*, /, +,%, &, ;)	&lnam..&dnam, where &vrnam < &dval
valid expressions	1 + 2, where age < 25

TABLE 2.3 Useful automatic SAS macro variables

KEYWORD	*BRIEF DESCRIPTION*
ALL, _AUTOMATIC_, _GLOBAL_, _LOCAL_, _USER_	All macro variables (users and automatic), all automatic macro variables, all global macro variables, all local macro variables, and all user-defined macro variable names and values can be displayed in the SAS log. Note that to display local macro variables, the %PUT _LOCAL_; statement needs to be placed within the macro.
SASEXECFILENAME	Name of SAS program executing. Note no '&' before the name.
&SQLOBS	Number of records selected from most recent PROC SQL.
&SYMEXIST	Confirms the existence of a macro variable.
&SYSDAY	SAS system day; for example, Monday.
&SYSDATE, &SYSDATE9	SAS system date; for example, 01JAN11, 01JAN2011.
&SYSDSN	Name of last SAS dataset accessed and libname as two words.
&SYSERR	0 if the last step was successful, else value > 0 for failed last step.
&SYSLAST	Name of last SAS dataset accessed with libname.
&SYSMACRONAME	Returns SAS macro name.
&SYSPBUFF	Stores all macro parameters passed once/PARMBUFF option is one when defining a macro program.
&SYSPARM	Character string that can be passed to SAS programs as macro parameter values when running in batch mode.
&SYSRC	Last return code from macro language execution, such as %SYSEXEC DIR *.SAS;.
&SYSTIME	SAS system time; for example, 10:34.
&SYSUSERID	User login ID.

TABLE 2.4 Useful debugging SAS macro options

KEYWORD	*BRIEF DESCRIPTION*
MERROR	Warning for unresolved macro calls.
MFILE	With MPRINT, useful to create macro free version of code.
MLOGIC, MLOGICNEST	Displays logical branches for each nested macro call and conditional execution, such as with &IF-%THEN. In version 9, use MLOGICNEST to display logical braches in all macro nested levels.
MPRINT, MPRINTNEST	Displays SAS code as macro is executed. In version 9, use MPRINTNEST to display SAS code in all macro nested levels.
MTRACE	Traces the execution of macros.
SERROR	Warning for unresolved macro variables.
SYMBOLGEN	Displays message when macro variable is resolved. Most useful for two or more ampersands. For example, &&DAT&I.

Note: See SAS paper on unraveling macros [22].

TABLE 2.5 Selected open code SAS macro statements

MACRO STATEMENT	*BRIEF DESCRIPTION*
%GLOBAL	Assign macro variables as global scope.
%LET	Assign global macro variables unless within a macro definition. Comparable to SYMPUT() and INTO: features during execution time.
%LOCAL	Assign macro variables as local scope.
%MACRO MYMAC; %MEND MYMAC;	Define macro program named MYMAC.
%MYMAC	Execute macro program named MYMAC.
%PUT	Display value of macro variables.
%SYMDEL	Delete global macro variables. For example, %SYMDEL libn dsn;.

Note: Selected open SAS macro statements such as %GLOBAL and %LET are compile time statements.

2.9 POWERFUL SAS MACRO QUOTING FUNCTIONS

SAS Macro Quoting Functions selection criteria to handle special characters (*, /, +, ;, <, >, ‘, “,), (, &, %) through a combination of several options: rescan or no rescan and compilation or execution:

	Rescan (Ignore spec chars)	**(NR) No Rescan (Ignore &, %)**
No pre-Compile	%str(proc print data = class; run;)	%nrstr(proc print data = **&dsname**; run;)
No pre-Execute		
Yes pre-Compile	%bquote(Summary of **N(#)** of AEs)	%nbquote(Summary of **N(%)** of AEs)
No pre-Execute	%bquote(‘Today’s News’)	
	%quote(123 O’Connor Street)	%nrquote(123 O’Connor **&Street**))
	%put I Live in %superq(**&state**).;	

1. Rescan macro variable to treat special characters, such as ‘;’, except ‘&’ and ‘%’ as literals, as in %str(proc print data = class; run;).
2. No Rescan macro variable to treat special characters, such as ‘;’ including ‘&’ and ‘%’ for macro variables and macro programs as literals, as in %nrstr(proc print data = &dsname; run;).
3. Do not precompile or preexecute by ignoring SAS syntax, such as %str(proc print data = class; run;).
4. Precompile to confirm valid macro and statement syntax. Do not preexecute for processing special characters in nonmacro syntax, such as %bquote (‘Today’s News’).
5. Prevents resolving macro variable names, such as in ‘I live in &state.’.

See the following tables for quick reference to SAS Macro Quoting Programming.

Table 2.6 Powerful SAS macro quoting functions
Table 2.7 Selected useful SAS macro data step call routines/functions/statements and macro functions

TABLE 2.6 Powerful SAS macro quoting functions

FUNCTION	*SAS FUNCTION BRIEF DESCRIPTION*
%BQUOTE()	Blind Quote — removes meaning of special characters, except '&' and '%', during macro execution. Useful for processing unbalanced quotes or parenthesis. Example: %BQUOTE(Summary of N (#) of Adverse Events),%BQUOTE('Today's News')
%NRBQUOTE()	No Rescan Blind Quote — removes meaning of special characters including '&' and '%' during macro execution. Example: %NRBQUOTE(Summary of N (%) of Adverse Events) %NRBQUOTE(^S = {font = ("arial", 11 pt)}Table 14-2. Demographics and Baseline Characteristics); &_textl = 'ABC Pharmaceuticals, Inc. ' || "%nrbquote(")" || ' j = r ' || "%nrbquote(")" || ' Page ~{pageof}'; &_textl = "ABC &_cpmstudyt" || "%nrbquote(")" || ' j = r ' || "%nrbquote(")" || ' Clinical Study Report';
%QUOTE()	Removes meaning of special characters, except '&' and '%'from a string during macro execution. Example: %QUOTE(123 O'Connor Street)
%NRQUOTE()	No Rescan Quote — removes meaning from a string including '&' and '%' during macro execution. Example: %NRQUOTE(123 O'Conner &Street)
%STR()	Removes meaning of special characters, except '&' and '%' at macro compilation. Example: %STR(PROC PRINT DATA = CLASS; RUN;)
%NRSTR()	No Rescan String — removes meaning of special characters including '&' and '%' at macro compilation. Example: %NRSTR(PROC PRINT DATA = &STUDY.. DEMOG; RUN;)
%SUPERQ()	Masks special characters including '&' and '%' at macro execution and prevents further resolution of the value. Example: %PUT I LIVE IN %SUPERQ(&STATE).;

Note: See SAS paper on quoting macro functions [11].

TABLE 2.7 Selected useful SAS macro data step call routines/functions/statements and macro functions

FUNCTION	*SAS FUNCTION BRIEF DESCRIPTION*
CALL EXECUTE	Allows various SAS code, such as PROC FREQ, to be executed in the middle of a Data Step.
CALL SYMPUT()	Data Step execution statement to create and populate macro variables with character values as compared to %LET to assign macro variables during compile time.
CALL SYMPUTN()	Data Step execution statement to create and populate macro variables with numeric values, as compared to %LET to assign macro variables during compile time.
%CMPRES()	Converts multiple blanks to single blanks. Note that this is comparable to the COMPBL() function and different from the COMPRESS() function which removes all blanks.
%DATATYP()	Returns 'NUMERIC' or 'CHAR' depending on whether the argument is an integer or a character string. Example: %DATATYP(35) will return 'NUMERIC' and %DATATYP(SERIOUS) will return 'CHAR'.
%DO; %END;	DO LOOP process similar to Data Step do loops for repeated execution.
%DO %TO %BY; %END;	DO-TO-BY-LOOP process similar to Data Step do-to-by-loops.
%DO %WHILE;	DO LOOP process similar to Data Step do loops for repeated execution.
%DO %UNTIL;	DO LOOP process similar to Data Step do loops for repeated execution.
%EVAL()	Preevaluate the integer macro expression as resolved value. See %SYSEVALF() for continuous number values.
%GOTO THERE; %THERE;	Transfer control to %THERE, labeled portion of the macro. Useful to exit macro if unexpected macro parameters for example.
%LEFT()	Shifts and removes leading blanks.
%LENGTH()	Calculated the length of the macro variable value. Zero length is missing value.
%LOWCASE()	Convert all text to lowercase values.
%IF %THEN; %ELSE;	IF-THEN-ELSE process similar to Data Step if-then-else.
%IF %THEN; %ELSE %IF %THEN; %ELSE %IF %THEN; ;	IF-THEN-ELSE-IF process similar to Data Step if-then-else-if. Note that an extra ';' is generally added at the end of the last ELSE-IF clause to execute the macro call or clause.
%INDEX()	Finds the first occurrence of a string.
%PUT	Displays text and macro variable values.
%QSCAN()	Scans the macro variable for words and returns quoted text. Useful to process list of values or keep macro variables unresolved.

(Continued)

TABLE 2.7 *(Continued)* Selected useful SAS macro data step call routines/functions/statements and macro functions

FUNCTION	*SAS FUNCTION BRIEF DESCRIPTION*
%QSUBSTR()	Selects text in macro variable based on string's position and returns quoted text.
%QSYSFUNC()	Preprocess macro value by first removing the meaning of special characters using internal SAS functions such as INPUTN().
%QUPCASE()	Convert all text values to uppercase values and returns quoted text.
RESOLVE()	Similar to the SYMGET() function, Data Step execution statement to access macro variables. In addition, the RESOLVE()function [23] can resolve macro and Data Step variables containing macro references at compilation (double quotes " ") or execution times (single quotes '').
%SCAN()	Scans the macro variable for words. Useful to process list of values.
%SUBSTR()	Selects text in macro variable based on string's position.
%SYMEXIST()	Returns 1 if macro variable exists else 0. Macro variable name is listed as argument without the '&' symbol.
%SYMGLOBL()	Returns 1 if macro variable is global else 0. Macro variable name is listed as argument without the '&' symbol.
%SYMLOCAL()	Returns 1 if macro variable is local else 0. Macro variable name is listed as argument without the '&' symbol.
%SYSCALL()	Access internal Data Step CALL routines such as DIR
%SYSEVALF()	Preevaluate the continuous number macro expression. See %EVAL() for integer values.
%SYSEXEC[]	Execute operating system commands immediately and save return code to &SYSRC global macro variable, such as %SYSEXEC DIR *.SAS;.
%SYSFUNC()	Preprocess macro value using internal SAS functions, such as INPUTN().
SYMGET()	Data Step execution statement to access macro variables and save as a character variable.
SYMGETN()	Data Step execution statement to access macro variables and save as a numeric variable.
%SYSGET()	Access values from macro variables or environmental variables, such as SAS_EXECFILENAME and SAS_EXECFILEPATH. Also supported by PROC SQL
%TRIM()	Removes trailing blanks.
%UNQUOTE()	Removes quotes.
%UPCASE()	Convert all text values to uppercase values.
%VERIFY()	Returns the position of the first character that is not in a text string which is the exact opposite to %INDEX(), Example: %VERIFY(123ABC, 1234567890);

REFERENCES

1. Wilson, Steven, The Validator: A Macro to Validate Parameters, SGF 2011. http://support.sas.com/resources/papers/proceedings11/015-2011.pdf.
2. Carpenter, Arthur, Advanced Macro Topics: Utilities and Examples, SUGI 23. http://www2.sas.com/proceedings/sugi23/Advtutor/p49.pdf.
3. How to pass in multiple values for a single parameter and then parse out each value. http://support.sas.com/kb/43/243.html.
4. %MACRO Statement. http://support.sas.com/documentation/cdl/en/mcrolref/61885/HTML/default/viewer.htm#macro-stmt.htm.
5. Accessing SAS System Information by Using DICTIONARY Tables. http://support.sas.com/documentation/cdl/en/sqlproc/62086/HTML/default/viewer.htm#a001385596.htm.
6. Dilorio, Frank, Data About Data: An Introduction to Dictionary Tables, SUGI 1995. http://www.sascommunity.org/sugi/SUGI95/Sugi-95-33%20Dilorio%20Michal.pdf.
7. Wilson, Steven, DEVELOPING A SAS SYSTEM AUTOCALL MACRO LIBRARY AS AN EFFECTIVE TOOLKIT, SUGI 1994. http://www.sascommunity.org/sugi/SUGI94/Sugi-94-15%20Wilson.pdf.
8. Carpenter, Arthur, Macro Functions: How to Make Them — How to Use Them, SUGI 27. http://www2.sas.com/proceedings/sugi27/p100-27.pdf.
9. Introduction to the Macro Facility. http://support.sas.com/onlinedoc/913/docMainpage.jsp?_topic = mcrolref.hlp/a002293969.htm.
10. Carpenter, Arthur, Using Macro Functions, SUGI 25. http://www2.sas.com/proceedings/sugi25/25/aa/25p004.pdf.
11. Carpenter, Arthur, Macro Quoting Functions, Other Special Character Masking Tools, and How To Use Them, SUGI 24. http://www2.sas.com/proceedings/sugi24/Advtutor/p38-24.pdf.
12. SAS Component Language Dictionary. http://support.sas.com/documentation/cdl/en/sclref/59578/HTML/default/viewer.htm#a000144382.htm.
13. Whitlock, Ian, CALL EXECUTE: How and Why, SUGI 22. http://www2.sas.com/proceedings/sugi22/CODERS/PAPER70.PDF.
14. Carpenter, Arthur, Kevin P. Delaney, SAS Macro: Symbols of Frustration?%Let us help! A Guide to Debugging Macros, SUGI 29. http://www2.sas.com/proceedings/sugi29/128-29.pdf.
15. Woodruff, Sarah, A Macro to Unravel Macros, SESUG 2009. http://analytics.ncsu.edu/sesug/2009/CC002.Woodruff.pdf.
16. Carpenter, Arthur, Storing and Using a List of Values in a Macro Variable, SUGI 30. http://www2.sas.com/proceedings/sugi30/028-30.pdf.
17. Clay, Ted, Tight Looping With Macro Arrays, SUGI 31. http://www2.sas.com/proceedings/sugi31/040-31.pdf.
18. Carpenter, Arthur, The Path, The Whole Path, And Nothing But the Path, So Help Me Windows, SGF 2008. http://www2.sas.com/proceedings/forum2008/023-2008.pdf.
19. Richardson, Kari, Eric Rossland, Using Macros to Automate SAS Processing, SUGI 29. http://www2.sas.com/proceedings/sugi29/126-29.pdf.
20. Whitlock, Ian, If and %IF You Don't Understand, SESUG 2010. http://analytics.ncsu.edu/sesug/2010/CC09.Whitlock.pdf.
21. Repole, Warren, Don't Be a SAS Dinosaur: Modernizing Programs with Base SAS 9.2 Enhancements, SGF 2009. http://support.sas.com/resources/papers/proceedings09/143-2009.pdf.

22. Johnson, Jim, Programming Squared (Writing Programs that Write Programs), NESUG 2001. http://www.nesug.org/proceedings/nesug01/at/at1011.pdf.
23. RESOLVE Function. http://support.sas.com/documentation/cdl/en/mcrolref/61885/HTML/default/viewer.htm#a000210258.htm.

CHAPTER 2: SAS MACRO PROGRAMMING QUESTIONS

1. Checking if a macro variable exists?
2. Creating macro variable to check if dataset exists?
3. Determining the number of records in a dataset?
4. Conditional processing if macro variable is populated?
5. A simple technique for scanning and creating a macro variable from a list of dataset names one at a time?
6. Which automatic macro variable can be used to identify the full path name of the SAS program?
7. Transferring control to another section of the program?
8. What are useful defensive programming techniques for checking the existence of external files, for example?
9. What is a useful SAS Version 9 function to create macro variables of numeric values using DATA step?
10. Which macro function is useful to dynamically create SAS statements using dataset values and then automatically executing the code?
11. In general, what are the three types of double ampersands (&&) or more for indirect references of macro variables?
12. What is one technique for accessing the dataset creation date, which is preserved instead of using the dataset file date which could be changed when the dataset is copied?
13. What options exist to demacrotize the SAS program to actual SAS statements?
14. Are there useful macros to clean up after SAS programs runs, such as clear libnames and delete temp formats, etc.?
15. What is the syntax for saving and compiling stored macro facility?
16. What are some of the key differences between the methods to create macro variables?
17. What statement is used to reset the SASMSTORE option?
18. Is there a limit to the number of characters in a macro call string?
19. Is there a useful general macro function that can accept almost any DATA step function?
20. What is the syntax for saving today's date as a date constant variable?
21. In general, what are differences in when to apply%IF/%THEN and IF/THEN conditions?
22. Is it possible to use the IN operator with macro conditions?

23. Are the corresponding four types of DO LOOP also available in macro programming?
24. What is an example of SAS macro programming without using any macro syntax?
25. What is the difference between %EVAL() and %SYSEVALF()?
26. What is the difference between positional and keyword macro parameters?
27. What are useful system options for debugging macro programs?
28. What are examples of macro statements that can be applied anywhere in the program or open code?
29. If a macro variable contains a comma and is used as a parameter in a macro call, then what is the correct method to prevent an error?

Advanced Programming Techniques

3

Chapter Overview

3.1 INTRODUCTION

This chapter is organized to facilitate easy searching of key points by summarizing and differentiating the syntax between similar SAS statements and options. Examples of SAS program and code statements are line numbered with references for more detailed explanation. Note that SAS examples are a complete block of code that can be executed, while selected SAS code syntax needs to be executed as part of the remaining program. This unique approach empowers both the advanced programmer who needs a quick refresher, as well as programmers interested in learning new programming techniques.

3.2 DEMONSTRATE THE USE OF ADVANCED TABLE LOOKUP TECHNIQUES

There are several methods for applying table lookup techniques. The basic objective is to replace or align values based on matching key values. In general, a variable containing old values needs to be replaced with new values, which may be stored in a second variable. The process of table lookup is to match based on the old value to replace or have the new value on the same record.

SAS TABLE LOOKUP TECHNIQUE	*RECORD ACCESS METHOD*	*COMMENTS*
IF-THEN/ELSE or CASE-WHEN/ ELSE for Table Lookup	Sequential access, transform one variable's set of old values to new set of values, update values within DATA step	Convenient, useful for conditional processing
Formats and PUT()/INPUT() for Table Lookups	Binary search to transform one character variable's set of old values to new set of values, update new values outside of DAT step and apply PUT() function and INPUT() function with informats for numeric values	Convenient, better than IF-THEN/ELSE method for character values. Generally fastest method for most lookups of under 30k to 40k items. Faster than MERGE and JOIN. Format catalog can be used to standardize searches and lookups across the organization.
Arrays and Do-Loops	Sequential access, **group related dataset variables or new variables or values to treat as single unit**	Convenient way for concise coding
SET with POINT= option and STOP statement	Direct access by record number, useful for referencing output datasets from SAS procedures such as PROC FREQ for DATA _Null_ reporting	Effective for small datasets or minimum number of record number references
DATA Step Merge	Join two or more tables by one or more **common key variables by exact matches**, requires pre-sorting datasets	Double SET statement merges can overcome come of the limitations of MERGE statement merge.
Proc SQL Joins	Join two or more tables by one or more key variables, does not require pre-sorting datasets, sometimes better than DATA Step merge for multiple-to-multiple joins, two datasets per join, multiple joins are needed for more than two datasets	Convenient way for concise coding, defaults for selecting variables is different from DATA step merge.
SAS Indexes	Direct access by key variable value for more effective ascending sorting/merging and subsetting, although PROC SORT is not required when using SAS indexes, applying PROC SORT makes it up to 50% faster	Up to 50% faster with PROC SORT if merging datasets, up to 100% slower without PROC SORT, may prevent other procedures such as PROC APPEND

Continued

SAS TABLE LOOKUP TECHNIQUE	*RECORD ACCESS METHOD*	*COMMENTS*
Hash Tables	Direct access by key variable value, **reference datasets within DATA steps as a multiple dimensional array but with both numeric and character values**	Has tables are temporary, up to 10% faster than SAS indexes

3.2.1 Array Processing

The real benefit of arrays is to make it easier to reference related new variables, dataset variables or a collection of values. Note that the same rules for variable names also apply within arrays. Arrays allow for buffer type processing. Arrays enable do-loop type processing to apply standard statements across variables. Arrays and do-loops work well together to create concise code. Note that array names can not contain numbers or special characters. Array statements are not executable.

Below are the different types of SAS arrays.

TYPES OF SAS ARRAYS	*SAS EXAMPLE*
One- or Multi-Dimensional, for multi-dimensional array, reference is number of rows then columns	One-Dimensional: array scr(4) s1 s2 s3 s4 (20 40 50 70); Multi-Dimensional: array resp(2,5) r1c1-r1c5 r2c1-r2c5; Above array is same as: array resp(2,5) r1c1 r1c2 r1c3 r1c4 r1c5 r2c1 r2c2 r2c3 r2c4 r2c5;
Numeric or Character arrays	Numeric: array scr(4) s1 s2 s3 s4 (20 40 50 70); Character: array chr(3) $ c1 c2 c3 ('x','y','z');
Variable list notation – one dash for suffix number, two dashes for root name, matching named prefix, all characters or all numeric variables in the dataset	array mis q1- q10; array atod a -- d; array vis (5) visit:; array days (*) _CHARACTER_; array sales(*) _NUMERIC_;
Implicit for SAS to determine array name, without () and number of elements, or Explicit arrays with array name()	Implicit: array mis q1 q2 q3 q4; Explicit: array sales(4) dept month year amount; array sales(*) dept month year amount;

Continued

TYPES OF SAS ARRAYS	*SAS EXAMPLE*
Permanent or Temporary arrays	Permanent: array scr(4) s1 s2 s3 s4 (20 40 50 70); * Remember to drop array variables – s1 s2 s3 s4; Temporary: array scr(4) _TEMPORARY_ s1 s2 s3 s4 (20 40 50 70); * Temporary arrays require dimension to be specified;
Array to create New Variables (useful for internal processing), to reference Dataset Variable Name or store Constant values in new variables	Array creates new variable names: array test(4); * test1 test2 test3 test4; Reference Dataset variable names: array sales(4) dept month year amount; Store Constant values in new variable names: array scr(4) s1 s2 s3 s4 (20 40 50 70); * Values are automatically retained;

One Array per Dataset variable, array number equals record number, Load dataset records into buffer for faster many-to-many merges, for example all males who match a female symp	data manymany2 (keep=f_patid m_patid f_symp m_diag f_diag rename=(f_symp=symp));
	
	* Use arrays to hold the retained (male) values, one array per variable – array name reference variable names, large number of elements for each record, array number is same as record number, temporary arrays, numeric or character arrays;
	array mpatids {100000} _temporary_;
	array mdiags {100000} $1 _temporary_;
	array msymps {100000} $2 _temporary_;
	
	* First Do loop to load dataset into array – keep selected variables and subset for males, for example;
	do until(done);
	set clindat.trial(keep=patid diag symp sex
	where=(sex='M'))
	end=done;
	
	* Save male data for diag and symp to three arrays;
	mcnt+1;
	mpatids{mcnt} = patid;
	mdiags{mcnt} = diag;
	msymps{mcnt} = symp;
	end;
	
	* Second Do loop to match with first dataset – rename same selected variables since same dataset and subset for females, for example;
	do until(fdone);
	set clindat.trial
	(rename=(patid=f_patid diag=f_diag symp=f_symp
	sex=f_sex)
	where=(f_sex='F')) end=fdone;

Continued

TYPES OF SAS ARRAYS	SAS EXAMPLE

	do i = 1 to mcnt; * 1 to total number of male records;
	 * Retrieve male values for matching female symp;
	 if msymps{i}=f_symp then do;
	  m_patid = mpatids{i}; * Male array references;
	  m_diag = mdiags{i};
	  output manymany2; * Save to dataset;
	 end;
	 end;
	 end;
	run;

Dynamic reference – DIM(), LBOUND() and HBOUND()	array sales(*) s1-s30;
	do J = 1 to dim(sales);
	 if sales(J) > 20000 then output;
	end;
	
	array sales(*) s1-s30;
	do J = lbound(sales) to hbound(sales);
	 if sales(J) > 20000 then output;
	end;

EXAMPLE – USEFUL TO GROUP RELATED VARIABLES	BRIEF DESCRIPTION
`array var(3) $ var1 - var3;`	Assign character variables to missing values using the array name as the base. var1=' '; var2=' '; var3 =' ';
`array goal{4} g1 - g4 (10, 15, 15, 10);`	Assign variables to numeric constant values with a list of variable names. g1 = 10; g2 = 15; g3 = 15; g4 = 10;
`array status{4} $ ('a', 'b', 'c', 'd');`	Assign variables to character constant values using the array name as the base. status1 = 'a'; status2 = 'b'; status3 = 'c'; status4 = 'd';
`array resp(2,5) r1c1-r1c5 r2c1-r2c5;`	Assign variable names to multi-dimensional array. r1c1=.; r1c2=.; r1c3=.; r1c4=.; r1c5=.; r2c1=.; r2c2=.; r2c3=.; r2c4=.; r2c5=.;
`array yrs(2,5) x1-10;`	Assign variable names to multi-dimensional array that is treated as a list of values that are similar to a one-dimensional array. Assigns x1 to yrs(1,1), x2 to yrs(1,2), x3 to yrs(1,3), x4 to yrs(1,4), and x5 to yrs(1,5) in row 1 and columns 1-5. In row 2, x6 to yrs(2,1), x7 to yrs(2,2), x8 to yrs(2,3), x9 to yrs(2,4), and x10 to yrs(2,5) in row 2 and columns 1-5.

3.2.2 Hash Objects

Hash objects or tables behave like super arrays and are often used within DATA null steps, but are not required. Hash objects can utilize indexes although the use of indexes is not needed. Three examples that show a) sort using one or more variables and variable selection, note that a hash sort is generally not more efficient than a PROC SORT, b) searching records and c) match-merge techniques (if and b) from one or more datasets. Hash tables prevent the need to pre-sort/index datasets, apply if/where conditions or merge/join datasets since all processing is done in memory. Hash tables are ideal for large datasets.

Four step process to create and apply Hash Tables:

1. Create a hash table with the DECLARE HASH statement.
2. Define the hash table with the DEFINEKEY(), DEFINEDATA() and DEFINEDONE() methods.
3. Fill the hash table with dataset values from DO-LOOP, SET and END= option. The hash object can also be loaded using the DATASET constructor.
4. A number of methods can be used to read from and write to the hash table with .OUTPUT() or .FIND(), for example.

Once DECLARE HASH <> is specified to name a hash object as well as one or more attributes, there are up to 26 hash methods that can be specified. One DECLARE HASH constructor is ORDERED: 'A', for example to specify ascending order. The DEFINEKEY() method is required to list one or more key numeric or character variables. With the DATASET constructor and a dataset name, the hash object is preloaded with all dataset values. This can be done with 'if _n_=1 then do; end;'. As an alternative method, the dataset attributes can be loaded in the PDV with 'if 0 then set <dataset name>';

Note that SAS dataset options can also be used while declaring and defining hash objects but it is limited to the following operations: RENAME, WHERE, DROP or KEEP. The DEFINEDATA() method lists all variables to save in the hash object followed by the DEFINEDONE() method. The alternative to using the DEFINEDATA() method is to specify a LENGTH statement with the selected variables. This process adds to the PDV. The DEFINEDONE() method is required. Output statement can be used in to specify an output dataset. Note that as an alternative, there is also an OUTPUT method.

Return codes enable conditional processing. A return code value of 0 means successful execution and no errors. Note that it is not required to reference hash objects with return codes, but it is a best practice as catastrophic failures have been reported when not used. Hash objects will execute as is.

General syntax: **HashSort**.DefineKey('**Title**')
Hash table name.Method("**Key Var**")

Return code reference enables conditional processing: rc= HashSort.DefineKey('Title')

Return codes used in conditional processing: if rc1=0 or rc2=0 then vitalstatus=0 else vitalstatus=1;

Best to use one of these three SAS examples to assure correct hash object syntax.

Example 1: Sort records and select variables from one SAS dataset

a. Load PDV and create sorted SAS dataset with selected variables

```
data _null_;
 if 0 then set movies;            * load variables into PDV;

 if _n_ = 1 then do;
  declare hash hashsort (ordered: 'a');       * required to create hash object;
  hashsort.definekey ('length');               * define key variable;
  hashsort.definedata ('title', 'length', 'category', 'rating');   * specify variables to keep;
  hashsort.definedone ();            * required end method;
 end;

 set movies end=eof;            * flag last record;

 hashsort.add ();                  * for each movie record add to hash object;
 if eof then hashsort.output(dataset: sorted_movies); * on last movie record, write
                                                        out to dataset;
run;
```

b. Load dataset in hash object and create sorted SAS dataset with selected variables

```
* Convert from original method of sorting with PROC SORT;
proc sort data=xmpl.marketing_rev out=xmpl.market_sort (keep=account_id contact_
  date);
 by account_id contact_date;
run;

* To sorting with hash objects to give the same result with less time;
data _null_;
 length account_id contact_date 8;

 if _n_=1 then do;
  declare hash hh (data set: 'xmpl.marketing_rev', ordered:'a');
  hh.definekey('account_id','contact_date');
  hh.definedone();
 end;

 hh.output(data set: ' xmpl.market_sort');
run;
```

c. Hash Tables can also reference data outside of SAS datasets

```
data highest_amt;
 length total_purchase_amount 8;

 set xmpl.account_names;
```

```
if _n_=1 then do;
 declare hash a(data set:'orcl.account_purchases');
 a.definekey('account_id');
 a.definedata('total_purchase_amount');
 a.definedone();
end;

if a.find() ge 0;
run;
```

Example 2: Load dataset in hash object and search records from another SAS dataset

```
/* Create Input Data Set */
data names;
 length first last title $ 16 born died 8;
 input first last born died title & $16.;
datalines;
William Blake 1757 1827 Spring
John Keats 1795 1821 To Autumn
Mary Shelley 1797 1851 Frankenstein
 ;
run;

/* Load and Find */
data _null_;
 length first last title $ 16;
 length born died 8;

 declare hash ht(dataset:"names");
  ht.defineKey("first", "last");
  ht.defineData("born", "died", "title");
  ht.defineDone();

/* Find John Keats */
 first = "John";
 last = "Keats";

 rc = ht.find();
 if rc = 0 then put "Found " first last title $QUOTE.;   * write one record to log;
else put "Not Found " first last;
run;
```

Output: Found John Keats "To Autumn"

Example 3: Match merge two SAS datasets (if a and b)

a. Merge two datasets by loading PDV from both datasets and output if match on key vairables

```
data match_on_movie_title(drop=rc);
 if 0 then set movies actors;                    *load variable into has tables;
```

```
 if _n_=1 then do;
  declare hash matchtitles(dataset: 'actors');          * create has table matchtitles;
  matchtitles.definekey('title');                       * define key variable;
  matchtitles.definedata('actor_leading', 'actor_supporting');  * specify variables to
                                                           keep from second
                                                           dataset;
  matchtitle.definedone();                              * complete hash table;
 end;

 set movies;

 if matchtitles.find(key;title)=0 then output;          * output if match title in
                                                          movies and actors;
run;              * keep all vars from movies and selected vars from actors;
```

b. Merge two datasets by loading PDV from both datasets and output if match on key vairables

```
data d_lookup (drop=rc);
 if 0 then set t_status new_status;                    * load variable properties into
                                                         hash tables;
 if _n_ = 1 then do;
  declare Hash MatchStatus (dataset:'new_status');     * declare the name MatchStatus
                                                         for hash;
  MatchStatus.DefineKey ('status');                    * identify variable to use as key;
  MatchStatus.DefineData ('new_status');               * identify columns of data;
  MatchStatus.DefineDone ();                           * complete hash table definition;
 end;

 set t_status;          * lookup status in t_status table using MatchStatus hash;

 if MatchStatus.find(key:status) = 0 then output;
run;
```

c. Merge two datasets by loading PDV and hash object from one dataset and output if match on key vairables

```
data hashmerge(keep=patid lname fname symp diag);
 if 0 then set clindat.patient;

 declare hash hmerge(dataset: 'clindat.patient', hashexp: 6);
  rc1 = hmerge.defineKey('patid');
  rc2 = hmerge.defineData('lname', 'fname');
  rc3 = hmerge.defineDone();

 do until(done);
  set clindat.trial end=done;
  rc4 = hmerge.find();
  if rc4 = 0 then output hashmerge;
 end;

 stop;
run;
```

Below is a list of twenty-six hash methods.

ADD() - Adds data associated with key to hash object.

CHECK() - Search for the key. If it is found, just return RC=0, and do nothing more. Note that calling this method does not overwrite the host variables.

CLEAR() - Removes all items from a hash object without deleting hash object.

DEFINEDATA() - Defines data to be stored in hash object. This method call can be omitted without harmful consequences if there is no need for non-key data in the table. Although a dummy call can still be issued, it is not required.

DEFINEDONE() - Tells SAS the definitions are done. If the DATASET argument is passed to the table's definition, load the table from the dataset.

DEFINEKEY() - Defines key variables to the hash object.

DELETE() - Deletes the hash or hash iterator object.

EQUALS() - Determines whether two hash objects are equal.

FIND() - Search for the key. If it is found, extract the satellite(s) from the table and update the host Data step variables.

FIND_NEXT() - The current list item in the key's multiple item list is set to the next item.

FIND_PREV() - The current list item in the key's multiple item list is set to the previous item.

FIRST() - Returns the first value in the hash object.

HAS_NEXT() - Determines whether another item is available in the current key's list.

HAS_PREV() - Determines whether a previous item is available in the current key's list.

LAST() - Returns the last value in the hash object.

NEXT() - Returns the next value in the hash object.

OUTPUT() - Dump the entire current contents of the table into a one or more SAS data set. Note that for the key(s) to dumped, they must be defined using the DEFINEDATA() method.

PREV() - Returns the previous value in the hash object.

REF() - Combines the FIND and ADD methods into a single method call.

REMOVE() - Removes the data associated with a key from the hash object.

REMOVEDUP() - Removes the data associated with a key's current data item from the hash object.

REPLACE() - If the key is not in the table, insert the key and its satellites, otherwise overwrite the satellites in the table for this key with new ones.

REPLACEDUP() - Replaces data associated with a key's current data item with new data.

SETCUR() - Specifies a starting key item for iteration.

SUM() - Retrieves a summary value for a given key from the hash table and stores the value to a DATA step variable.

SUMDUP() - Retrieves a summary value for the key's current data item and stores the value to a DATA step variable.

3.2.3 Formats

The use of PROC FORMAT to create formats used to map old values to new values is another example of the table lookup technique. The PUT() function is applied to convert the character variable.

```
proc format;
 value $status 'a' = 'aa'
               'b' = 'bb'
               'c' = 'cc'
               'd' = 'dd';
quit;
data d_lookup;
 set t_status;
  status = put(status, $status.);
run;
```

3.2.4 Combining/Merging Data

When merging two datasets by a common key variable, related information is joined at the record level. This table lookup technique is most effective for a large number of data values or batch processing. With this approach, the new values are on the same record as the old values.

```
proc sort data = t_status;
 by status;
run;
proc sort data = new_status;
 by status;
run;

data d_lookup (drop = status rename = (new_status = status));
 merge t_status (in = a) new_status;
 by status;
 if a;
run;
```

3.3 REDUCE I/O BY CONTROLLING THE SPACE REQUIRED TO STORE SAS DATASETS

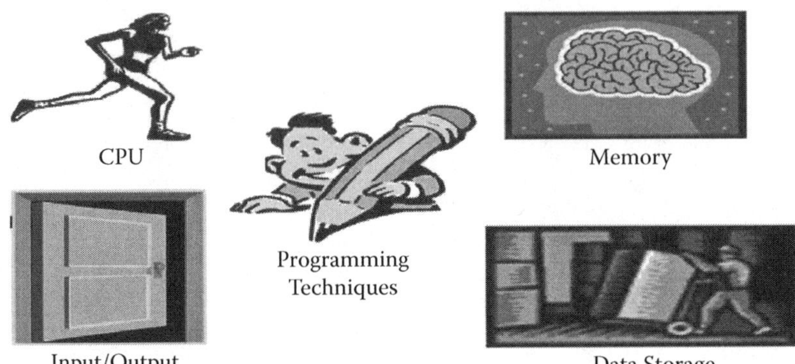

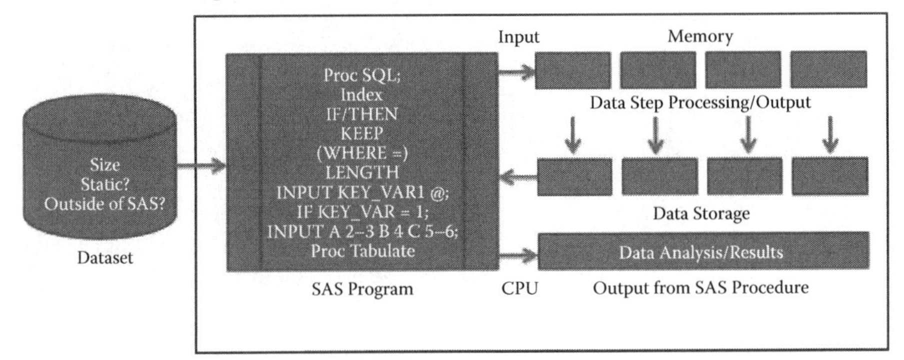

Often reducing input/output by controlling space requires balancing resources. For example, reducing CPU often results in reducing data storage and input/output but may increase memory usage. In general, 80% of the improvements can be expected using simple programming techniques. An additional 20% of the improvements can be expected with fine-tuning, optimal combination of resources, index, system options, hardware upgrades, and data warehouse strategies, such as summary-level datasets. See Figure 3.1. Factors (tradeoffs) to consider for SAS Program Efficiency table. See SAS paper on efficiency using SAS [2] [6] [12]. See SAS papers on system options [7] [8] [9].

Computer Processing Unit (Cpu)	Main Memory	Input/Output (I/O)	Data Storage Device	Programming Time
Measured in Seconds: Program time	**The maximum number of bytes of actual main memory the compilation and execution of the program used**	**Number of times a program requests a read/write operation between main memory and a data storage device**	**Physical amount of space required to store and process data sets**	**Time to develop, test, document and maintain programs**
- One vs. Multiple Pass? - Sorting/Indexing? Proc sort – cpu intense, physical order, requires 2.5 times size of input file Index – faster, logical order, need to be maintained, use on few variables, storage is about 10% of input file, best if data <u>does not change often</u> and expected to <u>select no more than 25% of all records</u> - Logic and priority conditioning to process fewer records? - Libname engines to directly access databases?	- Keep/Drop variables? - Summary/Partitioned/Snapshot data sets? - System option settings (BUFNO=, BUFSIZE=, MEMSIZE=, CATCACHE=, SORTSIZE 8m, NOMACRO, SASFILE) - SAS version 9 (THREAD option) - Assign large space to WORK library - Customize config or autoexec file.	- Keep/Drop variables? - Subset records using where/if? - Data set options/Data set statements? - Metadata info only to get variables and record counts – no need to read data set?	- Compress data sets to take less space? Most useful for short and wide data sets with character variables instead of long and narrow data sets. Can not be used with POINT= option. Requires SAS to compress & uncompress. - Variable length reduction, for example, to 3, for numeric variables? - Track stopping point when writing to large data sets? - Delete temporary data sets? - Use data view?	- Proc SQL to do multiple steps in one? - Check syntax only?-Min Proc Sort? - Class statement? - Format catalog? - Documentation? - Use random sample when possible? - Code for unknown and missing data values to avoid, for example, division by zero? - Assign descriptive and meaningful variable names? - Use the right SAS code for the task, for example to copy only the data set structure?

FIGURE 3.1 Factors (tradeoffs) to consider for SAS program efficiency: need to prioritize factors. Copyright © 2011 by Gupta Programming. See more SAS papers and references at: http://www.SASSavvy.com. Become a member today.

Below is a checklist of effective programming techniques:

____ Programming Techniques — using Proc SQL for multitasking
____ PROC DATASETS to modify dataset structure
____ CPU Time — using KEEP or DROP to limit number of variables
____ PROC SORT options, such as SORTEDBY = and PRESORTED options
____ Data Storage — using LENGTH to reduce variable size or SAS View instead of a SAS Dataset
____ Input/Output — using WHERE dataset option to limit number of records
____ Memory — reading only the records, variables, or datasets required
_ Use System Options, such as BUFNO =, BUFOBS =, BUFSIZE = COMPRESS =, etc.
_ Use DATA Step Options, such as NOMISS or NOSTMTID.
_ Use the LENGTH Statement to reduce the size of numeric variables and storage space.
_ Use numeric variables for analysis; otherwise, create character variables — less CPU intensive.
_ Use the KEEP or DROP statements to control only variables desired.
_ Delete unwanted datasets in the WORK area.
_ Combine steps to minimize the number of DATA or PROC steps.
_ Use data compression.
_ Conserve on memory (e.g., turning off NOMACRO, array processing)
_ Use Formats and Informats to save CPU during complex logic assignments.
_ Avoid unnecessary sorting with PROC SORT.
_ Control sorting by combining two or more variables at a time when sorting is necessary.
_ Use Subsetting IF statements to subset datasets.
_ Use WHERE statements to subset datasets.
_ Use indexes to optimize the retrieval of data.
_ Construct IF-THEN/ELSE statements to process condition(s) with greatest frequency first.
_ Save intermediate files in multistep applications.
_ Use PROC APPEND vs. SET statement to concatenate datasets.
_ Use PROC SQL to consolidate multiple operations into one step.
_ Use the PROC SQL Pass-Through Facility to pass logic to target database for processing.
_ Use the Stored Program Facility to store SAS DATA steps in a compiled format.
_ Use DATA Views and SQL Views to create "virtual" tables.
_ Use SAS Functions to perform common tasks.
_ Use the DATASETS Procedure COPY statement to copy datasets with built-in indexes.
_ Use the DATA _NULL_ step to avoid creating a dataset when one is not needed, but processing is.
_ Use a CLASS statement in procedures that support it to avoid having to sort data.

3.3.1 Compression Techniques

To improve data storage and input/output requirements, consider compressing large datasets. This technique is most useful for short and wide datasets with character variables instead of long and narrow datasets. Note that extra CPU is needed to uncompress datasets to be accessed again. In addition, compressed datasets cannot be used with the POINT = option.

```
data mylib.status (compress = yes);
 set status;
 < additional SAS statements >
run;
```

3.3.2 Length Statements

Reducing the length to the maximum character length instead of a much larger length will save space. This technique becomes more noticeable with a large number of character variables. Specify the LENGTH statement before the SET statement.

3.3.3 Eliminating Variables and Observations

By dropping unwanted variables and observations when first accessing the dataset, SAS programmers will save in having to process extra variables and observations. This will save in wasted time and space. Use the DROP or KEEP statement to keep only the required variables, and IF or WHERE statement to keep only the required observations. When subsetting datasets, the WHERE dataset or statement is more efficient than the IF statement. Note that the WHERE statement can be applied on SAS procedures also. Also, when possible, apply DROP and KEEP dataset options instead of SAS statements for faster response.

3.4 REDUCE PROGRAMMING TIME BY DEVELOPING REUSABLE SAS PROGRAMS

3.4.1 Data Step Views

Data step views store the instructions for viewing and accessing data. They do not take up storage space. One way to easily create SAS views is with PROC SQL. Instead of 'TABLE' in the PROC SQL syntax with the SELECT clause, 'VIEW' is specified. No other changes in the PROC SQL syntax are needed.

```
data myview view = myview;
 set mydata;
run;
```

```
proc sql;
 create view myview as select...;
quit;
```

3.4.2 DATA Steps That Write SAS Programs

See Chapter 2 SAS Macro Processing for examples of DATA steps that write SAS programs.

3.4.3 FCMP Procedure

The FCMP procedure allows you to create functions just like other SAS functions. The new functions can be compiled and used in the DATA step, the macro language, and within many SAS procedures that allow functions. See SAS papers on FCMP procedure [11] [12].

```
proc fcmp outlib = funcsol.functions.conversions;
 function cn2in(cn);
 in = cn/2.54;
 return (in);
 endsub;
run;

options cmplib = (funcsol.functions);

data inches;
 set sashelp.class(keep = name age height);
 inches = cn2in(height);
run;
```

3.5 PERFORM EFFECTIVE BENCHMARKING

3.5.1 Using the Appropriate SAS System Options

Useful system options for benchmarking program performance are: BUFNO =, BUFSIZE =, MEMSIZE =, CATCACHE =, SORTSIZE 8m, and SASFILE.

3.5.2 Interpreting the Resulting Resource Utilization Statistics

The concept benchmarking is to keep track of measurements, such as CPU and space, before and after making incremental changes in programming techniques and system options. In general, test cases that have optimal measurements help define the best programming approach to apply.

3.6 SORT PROCEDURE RESOURCES

3.6.1 Determine the Resources Used

When SAS sorts datasets, it takes both space and time. The space required is about 2.5 times the dataset space, since a new copy must be made. Once PROC SORT is executed, then the SAS log shows how much time it took to sort the dataset.

3.6.2 Avoid Unnecessary Sorts by Using Appropriate Indexes, Dataset Options, BY Statement Options, and the CLASS Statement

There are a number of SAS procedures that do not require datasets to be presorted. For example, PROC MEANS CLASS statement automatically sorts the dataset. See SAS paper on index [9].

3.7 IDENTIFY APPROPRIATE APPLICATIONS FOR USING INDEXES AND CREATE THEM

Indexes enable faster searches and subsetting. They need to be maintained and are best applied on a few selected variables. The extra index storage is about 10% of the input file. In addition, it is best if data does not change often, and expected subsetting results in no more than 25% of all records. The decision to create indexes or not is based on several factors. Since SAS indexes take up disk space and consume computer resources when being built and maintained, you will want to be selective about which variables you select as index key variables. Major factors to consider when creating indexes are if the following three criteria are met.

1. Which variable or variables will you most often use to subset the SAS dataset? Simple indexes use a single variable, and complex indexes use multiple variables.
2. Is your proposed index key variable discriminant?
3. Is the SAS dataset sorted into ascending order by the proposed index variable?

Once indexes are created, they are used in places such as:

- A WHERE statement in a DATA or PROC step
- A BY statement in a DATA or PROC step
- The KEY option on a SET statement
- The KEY option on a MODIFY statement

Nine step process to better understand and apply SAS indexes:

1. Instead of the normal sequential access method, implicit at the end of DATA step without an OUTPUT statement or explicit with OUTPUT statement, within a DATA step, SAS indexes offer a direct access method based on data values. The direct access method is similar to using the SET and POINT= option for direct accessing based on the record number. Similarly, the SET and KEY= option for direct accessing based on the indexed variable value is also within a do-loop to process all record and with the END=EOF to find the last record. When using the KEY= option, conditions can be applied on the _IORC_ temporary variable to check if =0. If _IORC_=0 then a record matching the key value was found. Note that for both methods for using SET and POINT= or KEY= option a useful technique is to apply a STOP statement to prevent an endless loop and is generally within a do-loop with the END=EOF option.
2. Once a SAS dataset is indexed, a separate index file is created and maintained. In addition, the index file becomes a property of the indexed SAS dataset. When not needed, SAS index files can be deleted as any other file or by using PROC SQL, for example. In addition, the SAS index file is deleted automatically if the original data set is changed and the SAS index is not re-created.
3. SAS indexes can potentially be helpful when repeatedly sorting, grouping, or subsetting a stable data table. Best to determine if SAS indexes are needed if one or more of these repetitive or transactional steps are generally done on one or more variables in large datasets: a. Sort by ascending order, b. Group by or c. Subset by. Note that SAS may decide not to use SAS indexes if it determines to be faster without SAS indexes.
4. A given index value should be no more than 10% of the data. Smaller percentages make for more efficiency. E.g. SEX makes a lousy index. In general, index variables should be unique or have limited duplicate values and be non-missing. Options include NOMISS to assure non-missing key variable and UNIQUE to assure unique key variables. Options are not required when creating SAS indexes. Applying an index stores the SAS dataset in ascending order by the key variable. This is required when using SAS indexes.
5. Which type of index to create, simple or complex: simple index is based on one variable and complex index is based on multiple variables. You can also have multiple indexes on a given table.
6. Your options for how to create SAS indexes are using a. DATA Step, b. PROC DATASETS or c. PROC SQL.
7. In the OPTIONS statement, make sure to specify MSGLEVEL=I to confirm SAS indexes are applied.
8. SAS applies SAS indexes in BY statements, for example, to faster process PROC SORT. SAS also applies SAS indexes when WHERE dataset option/statement or SET or MODIFY with KEY= option is specified. Note that SAS has some rules that must be met before applying SAS indexes. For example, if PROC SORT DESCENDING and NOTSORTED options are specified,

TABLE 3.1 Index subset guidelines

SUBSET SIZE	*INDEXING ACTION*
1%–15%	An index will definitely improve program performance
16%–20%	An index will probably improve program performance
21%–33%	An index might improve or it might worsen program performance
34%–100%	An index will not improve program performance

then SAS will not apply the SAS index. This means that SAS indexes do not work with the DESCENDING and NOTSORTED options. Note also that with an indexed SAS dataset, PROC SORT is not required before merging two datasets, however, applying PROC SORT makes the merge run faster. See Table 3.1 index subset guidelines for a quick reference.

9. When using PROC SQL on indexed SAS datasets, the syntax is unchanged and the process is generally faster.

3.7.1 Using the DATA Step to Create a Simple Index

```
data sales.bighits (index=(cdnumber / unique));
  set olddata.oldhits;
  < additional SAS statements >
run;
```

3.7.2 Using the DATA Step to Create a Complex Index

```
data sales.bighits (index=(numname=(cdnumber artistname)/nomiss));
  set olddata.oldhits;
  < additional SAS statements >
run;
```

3.7.3 Using PROC DATASETS to Create a Simple Index

```
proc datasets library=sales;
  modify bighits;
  index create cdnumber/unique;
run;
```

3.7.4 Using PROC DATASETS to Create a Complex Index

```
proc datasets library=sales;
  modify bighits;
  index create numname=(cdnumber artistname)/nomiss;
run;
```

3.7.5 Using PROC SQL to Create a Simple Index

```
proc sql;
  create unique index cdnumber on sales.bighits;
quit;
```

3.7.6 Using PROC SQL to Create a Complex Index

```
proc sql;
  create index numname on sales.bighits(cdnumber artistname);
quit;
```

3.7.7 Using PROC SQL to Delete a Simple Index

```
proc sql;
  drop index cdnumber from sales.bighits;
quit;
```

3.7.8 Using the DATA Step to Create and Use a Simple Index

```
options msglevel=i;
* class with index;
data work.class (index=(age));
  set sashelp.class;
run;

proc sort data= work.class;
  by age;
run;

* sequential access in ascending order to create a new non-indexed dataset;
data work.sortedclass;
  set work.class;
  by age;
run;
```

3.7.9 Using the DATA Step to Create and Use a Simple Index with the KEY= Option

```
* direct access for selected age values;
data work.direct;

  do age=13,14;
```

```
    do until (eof);
    set class key=age end=eof;
     if _IORC_=0 then do; /* 0 indicates a match was found */
       put _all_;
       output;
     end;
     else _ERROR_=0;    /* if no match, reset the error flag and continue */
    end;

   end;
   stop; /* stops future searching on ages 13 and 14 have been found */
run;
```

3.7.10 Using DATA Step to Create and Use a Simple Index with the KEY= Option with Values in Another Dataset

```
/* direct access */
data work.driver;
   age=13; output;
   age=14; output;
run;

data work.direct;

   set work.driver; /* <- sequential access & implicit loop */

   do until (eof);    /* <- explicit loop */
    set work.class key=age end=eof; /* <- direct access */
    if _IORC_=0 then do; /* 0 indicates a match was found */
      put _all_;
      output;
    end;
    else _ERROR_=0; /* if no match, reset the error flag and continue */
   end;
run;
```

3.8 COMPARE TECHNIQUES TO ELIMINATE DUPLICATE DATA

Table 3.2 briefly compares the general difference between the DATA step and PROC SQL. Selecting between using the DATA step or PROC SQL depends on several factors. See SAS paper comparing DATA Step and PROC SQL [2].

TABLE 3.2 DATA step vs. PROC SQL differences

In general, simple merges are faster with DATA steps, and complex merges are faster with PROC SQL. This is because DATA step uses sequential reads, while PROC SQL first creates a Cartesian Product and then subsets the dataset.
Dataset factors — size, static, relationships between datasets, density of matches, outside SAS access.
Operating system — Windows and Unix (generally similar), MVS and VMS (DATA step is generally faster).
One-stop report — PROC SQL is faster than 2 PROC Sorts, DATA merge, 1 PROC Sort, and 1 PROC Print.
Improve PROC SQL by up to 2X — When joining tables, process presorted tables as compared to unsorted tables.

3.8.1 Using the DATA Step

Using the FIRST. and LAST. temporary variables within the DATA step are useful to remove duplicate records. The dataset must first be presorted by the key variables. The code below saves duplicate records in the DUPS dataset and nonduplicate records in the UNIQUE dataset.

```
proc sort data = test;
 by id;
run;
data unique dups;
 set test;
 by id;
 if not(first.id and last.id) then output dups;
 else output unique;
run;
```

3.8.2 SORT Procedure

There are two options for removing duplicate records using PROC SORT. One option is to use NODUP with the BY variables, which may still leave duplicate records if the dataset is not sorted by the unique key variables. A better approach is to use the NODUPKEY with the BY _ALL_ reserved keyword to first sort by all variables before comparing each record.

```
proc sort data = sample nodupkey;
 by _all_;
run;
```

3.8.3 SQL Procedure

In general, PROC SQL is useful to identify duplicate records but not at removing duplicate records. This is because there is no FIRST. and LAST. Record-level processing

TABLE 3.3 PROC SQL advantages

Multitasking and sorting with one pass instead of multiple passes.		
SELECT name, sex	**(Memory) FROM sashelp.class**	(CPU with SORTEDBY = option when joining tables)
WHERE sex = 'F'	**(Input/Output)**	
GROUP BY sex	**(CPU)**	
HAVING weight > avg(weight)	**(Input/Output)**	
ORDER BY name;	**(CPU)**	
SQL pass-through for native database processing.		
Validate syntax before executing.		
Better programming style with concise and self-documenting code.		
(SORTEDBY=) [4] dataset option to take prevent resorting dataset.		
Utilize dictionary tables to access summary level data.		

Note: Insert table factors (tradeoffs) to consider for SAS program efficiency: from Chapter 3 landscape.doc file.

within PROC SQL as there is in DATA step programming. Also, with PROC SQL, each variable must be specified, as compared with the default of all variables being included in DATA step programming. Once duplicate records have been identified, however, other methods, such as DATA step programming or PROC SORT, can be applied to remove duplicate records.

```
proc sql;
 select id, count(*) as IDCount
 from test
 group by id;
quit;
```

Table 3.3 outlines the advantages of PROC SQL over DATA step programming. See SAS paper on efficiency techniques with PROC SQL [3].

REFERENCES

1. Lafler, Kirk Paul, An Introduction to SAS® Hash Programming Techniques, SESUG 2001. http://analytics.ncsu.edu/sesug/2011/BB08.Lafler.pdf.
2. Langston, Rick, Efficiency Considerations Using the SAS® System, SUGI 30. http://www2.sas.com/proceedings/sugi30/002-30.pdf.
3. Bhat, Gajanan, Raj Suligavi, Merging Tables in DATA Step vs. PROC SQL: Convenience and Efficiency Issues, SUGI 26. http://www2.sas.com/proceedings/sugi26/p104-26.pdf.
4. Stroupe, Jane, Linda Jolley, Dear Miss SASAnswers: A Guide to Efficient PROC SQL Coding, MWSUG 2009. http://www.lexjansen.com/mwsug/2009/saspres/MWSUG-2009-S06.pdf.

5. SORTEDBY = Data Set Option. http://support.sas.com/documentation/cdl/en/lrdict/64316/HTML/default/viewer.htm#a000131184.htm.
6. System Options Syntax Sorted Alphabetically. http://support.sas.com/documentation/cdl/en/allprodslang/63337/HTML/default/viewer.htm#syntaxByType-systemOption.htm.
7. Thornton, Patrick, Iuliana Barbalau, Working the System: Our Best SAS® Options, WUSS 2011. http://www.lexjansen.com/wuss/2011/applications/Papers_Thornton_P_74791.pdf.
8. Heaton, Edward, SAS® System Options are Your Friends, SESUG 2004. http://analytics.ncsu.edu/sesug/2004/TU07-Heaton.pdf.
9. Lafler, Kirk Paul, SAS® Performance Tuning Strategies and Techniques, SESUG 2003. http://www.scsug.org/SCSUGProceedings/2003/Lafler%20-%20SAS%20Performance%20Tuning%20Techniques.pdf.
10. Raithel, Michael, The Basics of Using SAS Indexes, SUGI 30. http://www2.sas.com/proceedings/sugi30/247-30.pdf.
11. Cohen, John J., Why Programming Efficiency Should Matter to All of Us, NESUG 2011. http://www.nesug.org/Proceedings/nesug11/ds/ds08.pdf.
12. Carpenter, Arthur, Using PROC FCMP to the Fullest: Getting Started and Doing More, MWSUG 2013. http://www.lexjansen.com/mwsug/2013/HW/MWSUG-2013-HW07.pdf.
13. Yves Deguire, Xiyun (Cheryl) Wang, Using SAS® PROC FCMP in SAS® System Development — Real Examples, SGF 2013. http://support.sas.com/resources/papers/proceedings13/505-2013.pdf.

CHAPTER 3: ADVANCED PROGRAMMING TECHNIQUES—QUESTIONS

1. How can you suppress notes and error messages?
2. What are some useful options for modifying the configuration file?
3. Does SASAUTOS automatically search subdirectories when a path is specified?
4. When submitting code in batch, is there a method to submit a line of syntax greater than 256 characters?
5. In general, what is a good matching records threshold value to use an index to increase efficiency?
6. What are at least three methods for applying table-look techniques?
7. What are the five main factors that determine a SAS program's efficiency?
8. In general, what percent improvement can be expected using simple programming techniques?
9. When applying the compression technique to save space, is it more effective for short and wide datasets or for long and narrow data sets?
10. In general, is it more efficient to apply DROP or KEEP as dataset options or SAS statements?
11. Do SAS views take up any space?
12. When using PROC SORT, how much disk space is required?
13. What are at least two methods for creating indexes?
14. Which two DATA step temporary variables are useful in identifying and removing duplicate records?

What's New in SAS Version 9.3

4

Chapter Overview

4.1 INTRODUCTION

This chapter is organized to facilitate easy searching of key points by summarizing and differentiating the syntax between similar SAS statements and options. Examples of SAS program and code statements are line numbered with references for more detailed explanation. Note that SAS examples are a complete block of code that can be executed, while selected SAS code syntax needs to be executed as part of the remaining program. This unique approach empowers both the advanced programmer who needs a quick refresher, as well as programmers interested in learning new programming techniques.

4.2 WHAT'S NEW IN SAS PROCEDURES

A new system option that enables the use of using more resources, if available, is THREADS. This option should help to reduce the processing time. Once turned on, many SAS procedures can then take advantage of it. See SAS reference [1] and SAS papers [2] [3].

Below is a list of selected SAS procedures with selected enhancements.

4.2.1 CIMPORT Procedure

PROC CIMPORT is used to convert transport files to SAS datasets. Below is a basic example.

```
Filename tranfile 'c:\temp\tran.xpt'; * Name of file to create;
Libname mydata '.';

*Option 1;
proc cport lib = mydata file = tranfile;
select times; * dataset name;
run;

*Option 2;
proc cport data = mydata.times file = tranfile;
run;
```

- — The CIMPORT SELECT and EXCLUDE statements support case sensitive names for files and catalogs in SAS/ACCESS engine libraries.
- — The CIMPORT procedure supports SAS name literals that include embedded blanks.
- — When VALIDVARNAME = ANY or VALIDMEMNAME = EXTEND are specified, the dataset names or member names used with the CIMPORT procedure can be up to 32 bytes in length. Names and member names can be mixed case.

4.2.2 CORR Procedure

PROC CORR is used to help quantify the relationship between two variables. Below is a basic example.

```
proc corr data = "D:\hsb2";
var read write math science female;
run;
```

- — The new POLYSERIAL option includes a table of polyserial correlation coefficients.

4.2.3 CPORT Procedure

PROC CPORT is used to convert SAS datasets to transport files. Below is a basic example.

```
Filename tranfile 'c:\temp\tran.xpt'; * Name of file to create;
Libname mydata '.';
* Option 1;
Filename tranfile 'c:\temp\tran.xpt'; * Name of file to restore;
Libname mydata '.';
```

```
Proc cimport infile = tranfile lib = mydata;
Select times; * dataset name;
Run;
* Option 2;
Proc cimport infile = tranfile data = mydata.times;
Run;
```

— The CPORT SELECT and EXCLUDE statements support case sensitive names for files and catalogs in SAS/ACCESS engine libraries.
— The CPORT procedure supports SAS name literals that include embedded blanks.
— When VALIDVARNAME = ANY or VALIDMEMNAME = EXTEND are specified, the dataset names or member names used with the CPORT procedure can be up to 32 bytes in length. Names and member names can be mixed case.

4.2.4 FCMP Procedure

The FCMP procedure allows you to create character or numeric functions just like other SAS functions. The new functions can be compiled and used in the DATA step, the macro language, and within many SAS procedures that allow functions. See SAS papers on FCMP procedure [5] [6]. Below is a basic example.

```
proc fcmp
outlib = work.funcs.Test; /* where will the functions be saved */
function whatAmI();       /* declare a function returning a number */
return(42);               /* return the number */
endsub;
function whereAmI() $;    /* declare a function returning a string */
return('In Test');        /* return the string */
endsub;
quit;
```

— INVCDF() computes the quantile from any distribution for which you have defined a cumulative distribution function (CDF).
— LIMMOMENT() computes the limited moment of any distribution for which you have defined a cumulative distribution function (CDF).

4.2.5 FORMAT Procedure

PROC FORMAT is useful for creating user-defined formats to display better and longer labels. Below is a basic example.

```
proc format library = myformat; /* Add sex format to format catalog */
 value $sex 'F' = 'Female'
            'M' = 'Male';
run;
```

— A user-defined format or informat that defines a missing value supersedes a value specified by the MISSING system option.
— The maximum number of labels that can be used for the MULTILABEL option is 255.
— The PICTURE statement directive %n formats the number of days in a duration.
— The PICTURE statement directive %s formats fractional seconds.
— Use the VALUE = statement to create a format that performs a function on a value.

4.2.6 The FREQ Procedure

PROC FREQ is useful for displaying frequency counts of both numeric and character variables. Below is a basic example.

```
proc freq data = Color;
  tables Eyes Hair Eyes*Hair/out = FreqCount outexpect sparse;
  weight Count;
run;
```

— Produces agreement plots when the AGREE option is specified and ODS graphics is enabled. In addition, also provides exact unconditional confidence limits for the relative risk and the risk difference.

4.2.7 PRINT Procedure

PROC PRINT is useful for displaying data values in a basic layout. Below is a basic example.

```
proc print data = exprev double;
   var country price sale_type;
run;
```

— Each BY group is a separate table, and the observation count is reset to zero at the beginning of each BY group.
— Bylines can be up to 512 characters.
— Except for the LISTING destination, if HEADING = V, the size of the column label is no longer restricted by the page size specified for the LISTING destination.
— For the LISTING destination, if HEADING = V, the variable name is used in place of a label if the column heading is too long for the page.
— ROWS = is valid only for the LISTING destination.
— If specifying a BY variable without first sorting on the BY variable, SAS stops printing and writes a message to the log.
— If the PRINT procedure errors or terminates, output might still be produced.

4.2.8 RANK Procedure

PROC RANK is useful to quantify variables by ranking their values. Below is a basic example.

```
proc rank data = test out = r_test;
 var spend;
run;
```

— The PRESERVERAWBYVALUES option preserves the raw values of the BY variable.

4.2.9 The REPORT Procedure

PROC REPORT is useful for creating publication quality detail and summary tables. Below is a basic example.

```
title '1.1a Continuous Data as a Summary Table';
proc report data = sashelp.class
 nowindows nocenter missing headline headskip nofs list split = '*';

 column (Sex,
     (('__ Age __'
      age = agen age = agemean age = agestd age = agemin age = agemax)
     ));

 define sex/across center;
 define agen/analysis n format = 3. 'N';
 define agemean/analysis mean format = 5.3 'Mean';
 define agestd/analysis std format = 5.3 'SD';
 define agemin/analysis min format = 3. 'Min';
 define agemax/analysis max format = 3. 'Max';
run;
```

— The MLF option has been added to the DEFINE statement in PROC REPORT.

4.2.10 SORT Procedure

PROC SORT is useful for sorting datasets. Below is a basic example.

```
proc sort data = employee;
  by idnumber;
run;
```

— The DATECOPY option copies to the output dataset the SAS internal date and time when the input dataset was created and the date and time when it was last modified prior to the sort.
— The new NOUNIQUEKEY, NOUNIQUEREC, and UNIQUEOUT = options have been added.

4.2.11 SQL Procedure

PROC SQL is useful for accessing and summarizing data, as well as creating basic listings. Below is a basic example.

```
proc sql;
SELECT name, sex FROM sashelp.class WHERE sex = 'F'
GROUP BY sex HAVING weight > avg(weight) ORDER BY name;
quit;
```

— Since SAS version 9.0, the INTO clause supports leading zeros as macro names when creating macro variables.

```
Select * into :n01 — :n10 from test; Creates 10 macro variables names n01, n02, ... n10;
```

4.2.12 TABULATE Procedure

PROC TABULATE is useful for summarizing data in table structure. Below is a basic example.

```
title1 '1a. Formatted Order, Categorical and Continuous Data as Columns';
title2 'Default of N and Sum';
proc tabulate data = sashelp.shoes missing;
 class region product subsidiary;
 keylabel n = 'Count' mean = 'Mean';
 var stores sales inventory returns;

 * Both character and numeric variables in display order;
 table product = 'Product' all = 'Total' sales = 'Sales' returns = 'Returns';

run;
```

— Available new statistics include upper and lower confidence limits, skewness, and kurtosis. In addition, ALPHA = option enables you to specify a confidence level.

— The NOCELLMERGE option has been added to the TABLE statement in PROC TABULATE.

4.2.13 UNIVARIATE Procedure

PROC UNIVARIATE is useful for providing descriptive statistics on continuous variables. Below is a basic example.

```
proc univariate data = BPressure;
  var Systolic Diastolic;
run;
```

— Supports five new fitted distributions: Gumbel distribution, inverse Gaussian distribution, generalized Pareto distribution, power function distribution, and Rayleigh distribution.
— These new distributions are available in the CDFPLOT, HISTOGRAM, PROBPLOT, PPPLOT, and QQPLOT statements.

4.3 WHAT'S NEW IN CHARACTER AND NUMERIC FUNCTIONS

The following functions and CALL routines are new:

- CALL RANCOMB() — permutes the values of the arguments, and returns a random combination of k out of n values.
- EFFRATE() — returns the effective annual interest rate.
- MVALID() — checks the validity of a character string for use as a SAS member name.
- NOMRATE() — returns the nominal annual interest rate.
- SAVINGS() — returns the balance of periodic savings by using variable interest rates.
- SQUANTILE() — returns the quantile from a distribution when you specify the right probability (SDF).
- SYSEXIST() — returns an indication of the existence of an operating environment variable.
- TIMEVALUE() — returns the equivalent of a reference amount at a base date by using variable interest rates.

Since SAS version 9.0, new character and numeric functions and formats allow greater flexibility in meeting programming requirements. There are more than 200 new functions in version 9.0. The following is a collection of new functions and informat: ANYALNUM, ANYALPHA, ANYDTDTEW(), CAT(), CATT(), CATS(), CATX(), COUNT(), COMPARE(), MEDIAN(), PCTL(), and SYMPUTX(). See SAS paper on functions [7].

ANYALNUM() — Returns the first position of alphanumeric character. This is helpful for variables storing alphanumeric values.

```
data _null_;
  string = 'abc 123 + =/';
  alpha = anyalnum(string);
  numeric = anyalnum(string, 5);
  other = anyalnum(string, 9);
  put alpha = numeric = other = ;
run;
/* alpha = 1 numeric = 5 other = 0 */
```

ANYALPHA() — Returns the first position of alphabetic character.

```
data _null_;
  string = 'abc 123 + =/';
  alpha = anyalpha(string);
  numeric = anyalpha(string, 5);
  other = anyalpha(string, 9);
  put alpha = numeric = other = ;
run;
/* alpha = 1 numeric = 0 other = 0 */
```

ANYDTDTEW() — This is an undocumented informat that is very helpful for reading dates in any valid date format. The user is not required to know the date format in advance. In addition, the date format can vary within the same date variable.

```
data dates;
 input mydate anydtdte9.;
 format mydate mmddyy10.;
 cards;         /* results */
14JAN1921  /* 01/14/1921 - ddmmmyyyy */
14011921   /* 01/14/1921 - ddmmyyyy */
01141921   /* 01/14/1921 - mmddyyyy */
;
run;
```

CAT(), CATT(), CATS(), CATX() — The family of CAT functions helps reduce the complexity of concatenating strings
CAT() — concatenate multiple strings
CATT() — CAT() plus TRIM()
CATS() — CAT() plus TRIM() plus LEFT()
CATX() — CAT() plus TRIM() plus LEFT() plus a Separator Delimiter
CATX() — concatenate multiple strings

```
data test;
 length a b $ 10;
 a = ' Good '; b = 'Afternoon';
 v9func = CATX('', a, b);            /* Good Afternoon */
 oldway = trim(left(a)) || ' ' || trim(b);
run;
```

COUNT() — The COUNT() function helps reduce the complexity of using the INDEX() and SUBSTR() functions to count the number of occurrences of a text within a string.

```
data test;
  a = 'Good Afternoon. Nice to be here.';
  x_oo = count(a, 'oo');          /* x_oo = 2 */
run;
```

COMPARE() — Returns the left-most character position when two strings differ or the value 0 if no difference exits.

Modifiers —

i or I ignores the case when comparing

l or L removes leading blanks before comparing

```
data test;
  infile datalines missover;
  input string1 $char8. string2 $char8. modifiers $char8.;
  result = compare(string1, string2, modifiers);
  datalines;                  /* results */
1234567812345678                /* 0 */
123      abc                  /* -1 */
abc      abx                  /* -3 */
aBc      AbC      i             /* 0 */
 abc     abc      l             /* 0 */
;
run;
```

MEDIAN() — Computes median value from list of nonmissing values.
  med = median(n1, n2, n3, n4);

```
obs  n1  n2  n3  n4  med
 1    1   2   3   4  2.5
```

PCTL() — Computes percentiles. pctvl = pctl(25, n1,n2,n3,n4);

```
obs  n1  n2  n3  n4  pctvl
 1    1   2   3   4  1.5
```

SYMPUTX() — Automatically converts numeric to character values and strips leading and trailing blanks.

```
data _null_;
  call symputx('v9way', 99.9);
  call symput('oldway', trim(left(put(99.9, 4.1))));
run;

%put &v9way &oldway;
99.9 99.9
```

4.4 WHAT'S NEW IN ODS STYLES AND TEMPLATES

The following are examples of ODS enhancements:

Now, the ODS graphics editor, the ODS graphics designer, and the ODS graphics procedures have moved from SAS/GRAPH to Base SAS. The Printer, PDF, PS, and PCL default printer values can now be changed in the SAS registry. Enhancements have been made to the DOCUMENT procedure, the TEMPLATE procedure, and the ODS statements.

In the DOCUMENT procedure, SAS/GRAPH external graph titles are now included in an ODS document, and the PRINT procedure is now fully supported.

In the TEMPLATE procedure, the default values for dynamic variables can now be supplied in the DYNAMIC, MVAR, and NMVAR statements for tabular output.

In ODS statements, such as ODS TAGSETS.RTF, these are examples of enhancements:

— OPTIONS (DOC = "changelog") provides version control information for the measured tagset.
— OPTIONS (TOC_LEVEL =) enables the user to define the number of levels that appear in the table of contents.
— OPTIONS (CONTINUE_TAG =) enables the user to add a continue tag to the RTF document when a table breaks and is continued to the next page.
— OPTIONS (WATERMARK =) enables the user to add a watermark to an RTF document. Use the ODS TAGSETS.RTF statement option WATERMARK to assign the text that will be displayed diagonally across each page of the RTF document.

See SAS website for more details [4].

4.5 WHAT'S NEW IN ODS STATISTICAL GRAPHICS

Now ODS statistical graphics is included with Base SAS. See SAS papers [8], [12], and [13]; ODS graph tip sheet [9]; and SAS galleries [10] and [11].

Table 4.1 summarizes PROC SGPLOT statements and their example plots.

Table 4.2 summarizes PROC SGPANEL statements and their example plots.

Table 4.3 summarizes PROC SGSCATTER statements and their example plots.

TABLE 4.1 Summary of PROC SGPLOT statements and their example plots

PLOT	*SYNTAX: PROC SGPLOT; <>; RUN;*
SCATTER	SCATTER X = var Y = var/options;
SERIES	SERIES X = var Y = var/options;
STEP	STEP X = var Y = var/options;
NEEDLE	NEEDLE X = var Y = var/options;
VECTOR	VECTOR X = var Y = var/options;
BUBBLE	Bubble X = var Y = var Size = var/options;
BAND	BAND X = var UPPER = var LOWER = var/options;
HIGHLOW	HIGHLOW X = var HIGH = var LOW = var/options;
REG	REG X = var Y = var/options;

(Continued)

TABLE 4.1 *(Continued)* Summary of PROC SGPLOT statements and their example plots

PLOT	*SYNTAX: PROC SGPLOT; <>; RUN;*
LOESS	LOESS X = var Y = var/options;
PBSPLINE	PBSPLINE X = var Y = var/options;
ELLIPSE	ELLIPSE X = var Y = var/options;
HBOX/VBOX	VBOX response-var/options; HBOX response-var/options;
HISTOGRAM	HISTOGRAM response-var/options;
DENSITY	DENSITY response-var/options;
HBAR/VBAR	VBAR category-var/options; HBAR category-var/options;
HLINE/VLINE	VLINE category-var/options; HLINE category-var/options;
DOT	DOT category-var/options;

(Continued)

TABLE 4.1 *(Continued)* Summary of PROC SGPLOT statements and their example plots

PLOT	*SYNTAX: PROC SGPLOT; <>; RUN;*
REFLINE	REFLINE value1 value2.../options;
XAXIS/YAXIS 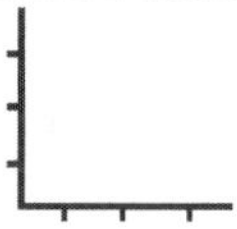	XAXIS options; YAXIS options;
INSET	INSET 'text1' 'text2'.../options;

TABLE 4.2 PROC SGPANEL statements and their example plots

PLOT	*SYNTAX: PROC SGPANEL; <>; RUN;*
Required PANELBY 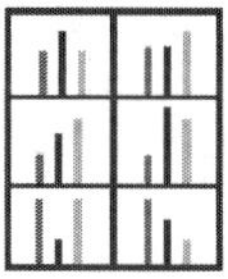	PANELBY var1 var2.../options;
Optional REFLINE 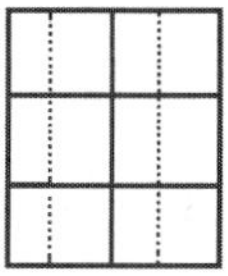	REFLINE value1 value2.../options;
COLAXIS/ ROWAXIS 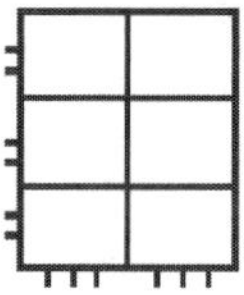	COLAXIS options; ROWAXIS options;

TABLE 4.3 PROC SGSCATTER statements and their example plots

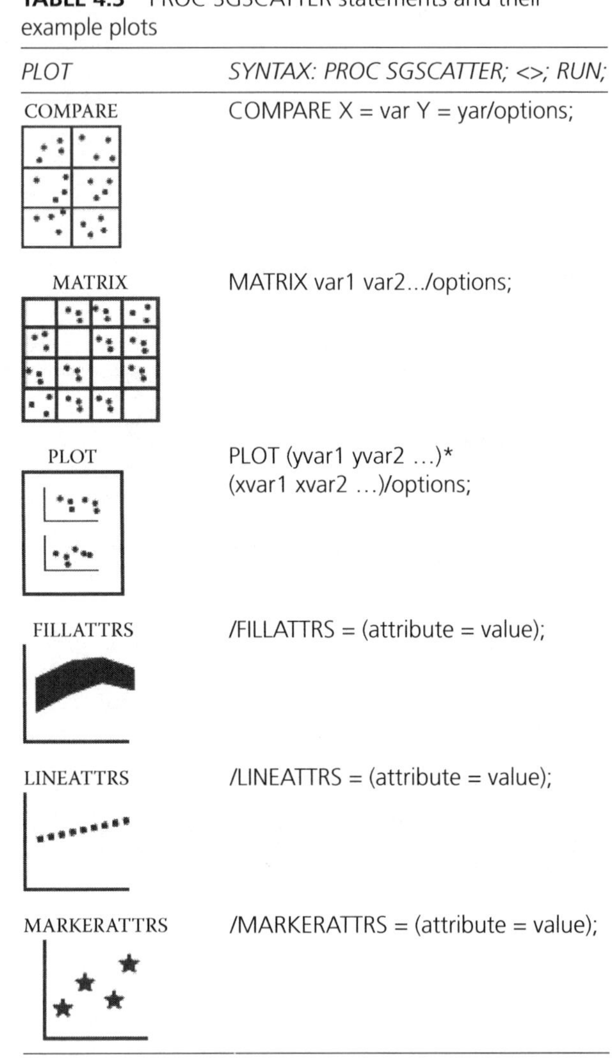

PLOT	*SYNTAX: PROC SGSCATTER; <>; RUN;*
COMPARE	COMPARE X = var Y = yar/options;
MATRIX	MATRIX var1 var2.../options;
PLOT	PLOT (yvar1 yvar2 ...)* (xvar1 xvar2 ...)/options;
FILLATTRS	/FILLATTRS = (attribute = value);
LINEATTRS	/LINEATTRS = (attribute = value);
MARKERATTRS	/MARKERATTRS = (attribute = value);

4.5.1 New Plot Options and Enhancements Are Available for the PROC SGPLOT, PROC SGPANEL, PROC SGRENDER, and PROC SGSCATTER Statements

— The ODS graphics procedures (SGDESIGN, SGPANEL, SGPLOT, SGRENDER, SGSCATTER), formerly called SAS/GRAPH statistical graphics procedures, are now part of base SAS software.

4.5.2 New Plot Statements for the SGPLOT and SGPANEL Procedures

— BUBBLE statement: A new BUBBLE statement creates a bubble plot in which two variables determine the location of the bubble centers and a third variable controls the size of the bubble.
— The new HBARPARM and VBARPARM statements create a horizontal or vertical bar chart based on a presummarized response value for each unique value of the category variable.
— The new HIGHLOW statement creates a display of floating vertical or horizontal lines or bars that represent high and low values. The statement also gives you the option to display open and close values as tick marks and to specify a variety of plot attributes.
— The new LINEPARM statement creates a straight line specified by a point and a slope. A single line can be created by specifying a constant for each required argument and multiple lines by specifying a numeric variable for any or all required arguments.

4.5.3 Updates to the PROC SGPLOT, PROC SGPANEL, and PROC SGSCATTER Statements

— The DATTRMAP = option specifies an SG attribute map dataset.
— The SGANNO = option specifies an SG annotation dataset.
— The PAD = option reserves space around the border of an annotated graph.
— The UNIFORM = option in the SGPLOT procedure enables you to control axis scaling and legend marker attributes for the row and column axes independently.

4.5.4 Updates to Plot Statements in the SGPLOT and SGPANEL Procedures

— The ATTRID = option specifies the value of the ID variable in an attribute map dataset.
— The CATEGORYORDER = option specifies the order in which the response values are arranged. This option affects bar charts, line plots, and dot plots.
— The CLIATTRS = and CLMATTRS = options enable you to specify line attributes and fill attributes for confidence limits.
— The CURVELABELATTRS = and DATALABELATTRS = options specify options for setting text attributes for plot curves and labels.
— The DISCRETEOFFSET = option specifies an amount to offset graph elements from the category midpoints or from the discrete axis tick marks. This option affects bar charts and box plots.

— The following are new options for grouped data (using the GROUP = option):
The CLUSTERWIDTH = option specifies the cluster width as a ratio of the midpoint spacing when a group is in effect. This option affects any plot that can have a discrete axis.
The GROUPDISPLAY = option specifies how to display grouped graphics elements. This option affects any plot that can have a discrete axis. (The option is not available for the HBARPARM and VBARPARM statements.)
The GROUPORDER = option specifies the ordering of graph elements within a group. This option affects any plot that can have a discrete axis.

4.5.5 BAND Statement

— The CURVELABELLOWER = and CURVELABELUPPER = options specify labels for the plot's upper and lower limits.
— The TYPE = option specifies whether the data points for the band boundaries are connected as a series plot or as a step plot.

4.5.6 HBAR and VBAR Statements

— The DATALABEL = option now enables you to specify a variable that contains values for the data labels.
— The DATASKIN = option specifies a special effect to be used on all filled bars.
— Some SAS styles display fill patterns for grouped bars.

4.5.7 HBOX and VBOX Statements

— The CAPSHAPE = option specifies the shape of the whisker cap lines.
— The CONNECT = option specifies that a connect line joins a statistic from box to box.
— Boxes can be grouped. Along with the GROUP = option, the GROUPDISPLAY = and GROUPORDER = options are available.
— The NOTCHES option shows the notches.
— The NOMEAN option hides the mean symbol.
— The NOMEDIAN option hides the median line.
— The NOOUTLIERS option hides the outliers.
— You can specify attributes for these elements: connect lines, data labels, box fills and lines, mean markers, median lines, outlier markers, and whisker and cap lines.

4.5.8 HISTOGRAM Statement

— BINSTART = specifies the X coordinate of the first bin.
— BINWIDTH = specifies the bin width.

— NBINS = specifies the number of bins.
— The TITLEATTRS = and TEXTATTRS = options in the INSET statement in the SGPLOT procedure.
— The TITLEATTRS = and VALUEATTRS = options in the KEYLEGEND statement.

4.5.9 VLINE Statement

— The DATALABELPOS = option specifies the location of the data label.

4.5.10 Axis Updates for the SGPLOT Procedure

— The XAXIS, X2AXIS, YAXIS, and Y2AXIS statements support several enhancements and new options:
New LABELATTRS and VALUEATTRS options specify textual attributes for axis labels and axis tick value labels, respectively.
— A new REVERSE option specifies that the tick values are displayed in reverse (descending) order.
— New THRESHOLDMAX and THRESHOLDMIN options specify a threshold for displaying one more tick mark at the high end and the low end of the axis, respectively.

4.5.11 Axis Updates for the SGPANEL Procedure

— Same axis updates for the SGPLOT procedure.
— The REFTICKS option enables you to specify whether labels and values are added to the tick marks. This option adds tick marks to the side of the panel that is opposite from the specified axis.

REFERENCES

1. What's New in SAS 9.3. http://support.sas.com/documentation/cdl/en/whatsnew/64209/PDF/default/whatsnew.pdf.
2. Fecht, Marje, SAS 9 Programming Enhancements, SESUG 2005. http://analytics.ncsu.edu/sesug/2005/HW02_05.PDF.
3. Repole, Warren, Don't Be a SAS Dinosaur: Modernizing Programs with Base SAS 9.2 Enhancements, SGF 2009. http://support.sas.com/resources/papers/proceedings09/143-2009.pdf.
4. What's New in the Output Delivery System. http://support.sas.com/documentation/cdl/en/odsug/65308/HTML/default/viewer.htm#odsugwhatsnew93.htm.
5. Carpenter, Arthur, Using PROC FCMP to the Fullest: Getting Started and Doing More, MWSUG 2013. http://www.lexjansen.com/mwsug/2013/HW/MWSUG-2013-HW07.pdf.

6. Yves Deguire, Xiyun (Cheryl) Wang, Using SAS® PROC FCMP in SAS® System Development — Real Examples, SGF 2013. http://support.sas.com/resources/papers/proceedings13/505-2013.pdf.
7. Gupta, Sunil, Getting Familiar with SAS® Version 8.2 and 9.0 Enhancements, WUSS 2003. http://www.lexjansen.com/wuss/2003/Tutorials/i-getting_familiar_with_sas_version_8_and_9.pdf.
8. Slaughter, Susan, Lora D. Delwiche, Graphing Made Easy with SG Procedures, SGF 2012. http://support.sas.com/resources/papers/proceedings12/259-2012.pdf.
9. ODS Graph Tip sheet. http://support.sas.com/rnd/app/ODSGraphics/TipSheet_ODSGraphics.pdf.
10. Graphics Samples Output Gallery. http://support.sas.com/sassamples/graphgallery/index.html.
11. Graphically Speaking Blog, Visual Index. http://support.sas.com/rnd/datavisualization/graphicallyspeakingindex/.
12. Kincaid, Chuck, Using the 9.2 Statistical Graphic Procedures, NESUG 2011. http://www.nesug.org/Proceedings/nesug11/hw/hw02.pdf.
13. Delaney, Kevin P., Multiple Graphs on One Page, the easy way (PDF) and the hard way (RTF). http://www2.sas.com/proceedings/sugi28/094-28.pdf.

CHAPTER 4: WHAT'S NEW IN SAS VERSION 9.3—QUESTIONS

1. What is the PROC SORT option to make sorts case insensitive?
2. Which function is useful for counting words in a string?
3. Which function is useful for locating a word and word number?
4. Which function is useful for extracting the first character in a string?
5. What is the difference between CPORT and CIMPORT?
6. Which CIMPORT options will enable dataset names or member names to be up to 32 bytes in length?
7. Can PROC FCMP be used to create character or numeric functions?
8. What is the option that copies to the output dataset the SAS internal date and time when the input dataset was created and the date and time when it was last modified prior to the sort?
9. Which function returns the first position of alphanumeric character?
10. Which new SAS 9.0 function is similar to CAT() plus TRIM() plus LEFT()?

Answers to Chapter Review Questions

5

CHAPTER 1: ACCESSING DATA USING SQL—ANSWERS

1. **Creating a nonmissing variable from multiple datasets?** Use COALESCE() function.
2. **Creating character string macro variables without any blanks?** Use !! and TRIM() function.
3. **Creating an empty dataset with minimum code and no uninitialized notes?** Use PROC SQL; CREATE TABLE SHELL (NAME CHAR, AGE NUM); QUIT;
4. **What technique is useful for identifying and adding baseline lab values in clinical trials?** Generally, the baseline visit is defined as the last visit before the first dose date. PROC SQL; CREATE TABLE _BRES AS SELECT LBTESTID, SUBJID, RESULT AS BRESSAFE FROM QC_ALAB_RAW2 WHERE RESULT >. AND VISITDTF < FDOSEDTF GROUP BY SUBJID, LBTESTID HAVING VISITDTF = MAX(VISITDTF); CREATE TABLE QC_ALAB_RAW3 AS SELECT A.*, B.* FROM QC_ALAB_RAW2 AS A FULL JOIN _BRES AS B ON A.SUBJID = B.SUBJID AND A.LBTESTID = B.LBTESTID ; QUIT;
5. **In general, when having nested summary functions, such as SUM(SUM(VOL1, VOL2)*10) with a GROUP BY clause, what is the difference in the two SUM() functions?** The inner SUM() is first applied to total VOL1 and VOL2 at the record level times 10, then the outer SUM() is applied to total all of the individual records to get one record per GROUP BY variables.
6. **When applying a condition to subset the first dataset A used in a LEFT JOIN with a second dataset B, does PROC SQL subset based on the condition after the join or apply the LEFT JOIN to keep all records in dataset A?**

PROC SQL keeps all records in dataset A and ignores the condition on dataset A. As an option, apply a WHERE dataset option to first subset dataset A.

7. **Is it possible to select all variables in a dataset as well as apply the COALESCE() function to keep nonmissing values in the same SELECT statement when joining two tables?** In general, yes; make sure to first apply the COALESCE() function before selecting all variables to correctly define the possible nonmissing variable. Switching the order may not work. SELECT UNIQUE COALESCE(a.dosegrp, c.dosegrp) AS dosegrp, A.* FROM conmeds_all2 AS A LEFT JOIN analyvar AS C ON A.subjid = C.subjid ORDER BY dose, scrid; As an alternative to preventing the WARNING message, save to a new variable name, such as B_DOSEGRP, and then apply DROP DOSEGRP and RENAME B-DOSEGRP to DOSEGRP as dataset options.
8. **What PROC SQL option can be added to prevent any warnings?** NOWARN option will prevent any warning messages. This is helpful when using B.*, which adds common key variable names, for example. Make sure confirm results before turning off warning message.
9. **Is it possible to control and loop through each record in a dataset as done with the _N_ in the DATA step?** Yes. With the WHERE MONOTONIC() = &I clause, you can insert PROC SQL within a do loop to increment the value of I.
10. **In general, do variables listed in the GROUP BY clause need to be included in the SELECT clause?** Yes, in general, you need to include, in the same order, the same variables in the SELECT as you specify in the GROUP BY clause. As an alternative, you can presort the dataset with a variable you do not want in the PROC SQL to control the sort order. This technique is helpful to sort the dataset based on a detail variable, and then to create one summary record to merge back without the detail variable. As needed, you can then include new variables in the SELECT clause.
11. **When submitting PROC SQL code in batch mode, is there a method to submit a line of syntax greater than 256 characters?** Yes, by including line breaks to break up the 256-character statement. Another option is to save the long PROC SQL line of code in a separate file and %INCLUDE the code and set the LRECL = option as in this example — FILENAME MYFILE 'PATH TO FILE' LRECL = 32767;
12. **Is it possible to select more variables when applying EXCEPT or INTERSECT joins?** Yes, since EXCEPT or INTERSECT selects limited key variables to identify specific records, you can apply the EXCEPT or INTERSECT SELECT as a subquery that applies condition when selecting other variables.
13. **What are some techniques to prevent this 'NOTE: The query requires remerging summary statistics back with the original data'?** If you use a summary function in a SELECT clause or a HAVING clause without a GROUP BY clause then you will get this note in the SAS log. This message is only a NOTE which states that PROC SQL must remerge data (or make two passes through the table). The first pass calculates and returns the summary function values and the second pass retrieves all other variables in the SELECT clause.

Remerging just requires additional processing time and is often unavoidable. This is because the summary function creates overall summary stats based on all records and is not by a group variable which is generally used. Grand percentages, however, are common without GROUP BY clause.

14. **Is it possible to calculate event percentages such as total # of patients with events/total # of patients using PROC SQL?** Yes, see the e-guide for an example with the DATA step. Additional useful information is total number of events.
15. **What is one advantage that PROC SQL has over DATA step when joining datasets?** PROC SQL allows for joining by different variable names instead of requiring the same variable name.
16. **What is the syntax to apply the colon modifier ':' to match based on values only and exclude blanks?** In PROC SQL, use NAME EQT 'Sunil';
17. **What is the PROC SQL syntax to create macro variables?** PROC SQL; SELECT count(DEPT) INTO :DEPTCNT; QUIT;
18. **When doing a many-to-many join without a WHERE clause, what is the danger?** Without a WHERE clause causes unrelated records to be linked from both datasets and creates a Cartesian product. With the WHERE clause to link related records, each record per key variable is linked with each record in the second dataset. The correct total number of records is the number of records per key variable in the first dataset times the total number of records per key variable in the second dataset. Using the DATA step with or without the WHERE statement may cause incorrect joins and incorrect number of records. The number of records created is based on the maximum number of records per key variable from the datasets.
19. **Is there a method to pull data from multiple datasets and create flag variables for non-missing values to get patient accountability?** Yes, first create a population dataset by creating flag variables using IFN(AGE > ., 1, .) and then LEFT JOIN with the population dataset to create a summary dataset with all selected variables to be used in PROC PRINT. PROC FREQ is used to display all flag variable combinations and PROC PRINT with WHERE to display detail records.

CHAPTER 2: SAS MACRO PROGRAMMING—ANSWERS

1. **Checking if a macro variable exists?** %if %sysfunc(symglobal(<macro_name>)) = 1 %then %do; or %if %sysfunc(symlocal(<macro_name>)) = 1 %then %do;
2. **Creating macro variable to check if dataset exists?** Use %let dsetyn = %sysfunc(exist(<dataset_name>));

3. **Determining the number of records in a dataset?** Can use &SQLOBS from most recent PROC SQL SELECT statement.
4. **Conditional processing if macro variable is populated?** Can use %if %length(<macro_variable>) > 0 %then %do; %end;
5. **A simple technique for scanning and creating a macro variable from a list of dataset names one at a time?** Can use this code: %do dupi = 1 %to 3; %let dupdsn = %scan(DATASET1 DATASET2 DATASET3, &dupi); %end;
6. **Which automatic macro variable can be used to identify the full path name of the SAS program?**
   Can use this macro:
   %macro prgpath;
   %qsubstr(%sysget(SAS_EXECFILEPATH), 1, %length(%sysget(SAS_EXECFILEPATH))- %length(%sysget(SAS_EXECFILEname)))
   %mend prgpath;
   %let libpath = %qsubstr(%prgpath,1,55)&pgm._toc.rtf;
   %put &libpath;
7. **Transferring control to another section of the program?** Can use %GOTO THERE; to direct program to %THERE: position.
8. **What are useful defensive programming techniques for checking the existence of external files, for example?** %IF %SYSFUNC(FILEEXIST(&OBJECT_NAME)) = 1 %THEN %DO;
9. **What is a useful SAS Version 9 function to create macro variables of numeric values using DATA step?** SYMPUTX(dsvar1, dsvar2) automatically applies TRIM(LEFT(PUT())) to the numeric value.
10. **Which macro function is useful to dynamically create SAS statements using data set values and then automatically executing the code?** CALL EXECUTE() also is useful to conditionally execute macro definitions.
11. **In general, what are the three types of double ampersands (&&) or more for indirect references of macro variables?** Use &&&XX to reference one macro variable. Use &&XX& to reference a list of root word multiple variables, such as &dsn1, &dsn2, and &dsn3. Use &&&XX& to reference a list of any multiple variables, such as &engine1, &engine2, &mpg1, and &mpg2.
12. **What is one technique for accessing the dataset creation date that is preserved instead of using the dataset file date, which could be changed when the dataset is copied?** %let download = %substr(%sysfunc(attrn(&dsid,crdte),datetime.),1,7);
13. **What options exist to demacrotize the SAS program to actual SAS statements?** MFILE option.
14. **Are there useful macros to clean up after SAS programs runs, such as clear libnames and delete temp formats, etc.?** Yes, you can apply LIBNAME, PROC CATALOG for example. Clean SAS programming environment.
15. **What is the syntax for saving and compiling stored macro facility?** Apply the STORE and SOURCE options when creating the macro and the MSTORED option to access the macro. See macro documentation for more info (search for store macro facility, select Storing and Reusing Macros).

16. **What are some of the key differences between the methods to create macro variables?** %LET, SYMPUT, and INTO.
17. **What statement is used to reset the SASMSTORE option?** %SASMSTORECLEAR;
18. **Is there a limit to the number of characters in a macro call string?** No, can include line breaks to break up the 256-character macro call. Another method is to save the macro call in a separate file and INCLUDE the macro call program. See MVARSIZE system option.
19. **Is there a useful general macro function that can accept almost any DATA step function?** Yes, the %SYSFUNC() macro function can accept almost any DATA step function. This is ideal for nonmacro functions, such as %SYSFUNC(STRIP(X)).
20. **What is the syntax for saving today's date as a date constant variable?** %let dtcutoff = &sysdate; %put Todays Date = &dtcutoff; data visit_qc; set dbs.visit; dtcutoff = "&dtcutoff"d; format dtcutoff date9.; if visendt >. and dtcutoff >. then studydy = dtcutoff - datepart(visendt); else studydy =.; run;
21. **In general, what are differences in when to apply %IF/%THEN and IF/THEN conditions?** %IF/%THEN conditions become true only when user or system macro variables equal a constant, for example, while IF/THEN conditions become true when both user/system macro variables or dataset variables equal a constant. This means that %IF/%THEN condition determines which SAS block of code gets compiled, while IF/THEN condition determines which SAS block of code is executed. All of the IF/THEN block of code is compiled. Note that numeric and character values are treated the same; quotes are not applied.
22. **Is it possible to use the IN operator with macro conditions?** Yes, with the MINOPERATOR and MINDELIMITER options. Note that there are no quotes with IN.
23. **Are the corresponding four types of DO LOOP also available in macro programming?** Yes, all four types are available - %DO; %END;, %DO %TO %BY; %END;, %DO %WHILE;, %DO %UNTIL;
24. **What is an example of SAS macro programming without using any macro syntax?** Within a DATA step, using arrays and do-loops to repeat a block of SAS code for different variables or values resembles SAS macro programming.
25. **What is the difference between %EVAL() and %SYSEVALF()?** Both %EVAL and %SYSEVALF() evaluate the math expression before assigning it as a macro value. %EVAL() is for integers and %SYSEVALF() is for continuous values. ex. %LET RESULT = %EVAL(&NUMER/&DENOM);
26. **What is the difference between positional and keyword macro parameters?** Keyword parameters are ideal for setting default values, additional flexibility in specifying macro variables in any order and better documentation. Positional parameters must be called in the exact same order as defined.
27. **What are useful system options for debugging macro programs?** MPRINT (log shows the code actually created by macro), MLOGIC (log shows flow of macro conditional execution), MPRINTNEST (shows macro

nesting such as MPRINT(OUTER.INNER.INRMOST): 'This is the text of the PUT statement') MERROR and SERROR to display warnings, and SYMBOLGEN (log shows macro variable value). MPRINT and MFILE options can be used to create a macro free version of the code. In addition, the automatic macro variables _ALL_, _LOCAL_, _GLOBAL_, _USER_, or _AUTOMATIC_ can be displayed.

28. **What are examples of macro statements that can be applied anywhere in the program or open code?** %LET, %PUT along with many macro functions such as %INDEX(), %SCAN() and %LEFT().
29. **If a macro variable contains a comma and is used as a parameter in a macro call, then what is the correct method to prevent an error?** Use the %BQUOTE() to mask the commas as in %TEST(%BQUOTE(&VAR), B, C);

CHAPTER 3: ADVANCED PROGRAMMING TECHNIQUES—ANSWERS

1. **How can you suppress notes and error messages?** Use OPTIONS NONOTES ERRORS = 0; Warnings cannot be suppressed. Remember to reset option to display notes.
2. **What are some useful options for modifying the configuration file?** WORK, REGISTER, ALTLOG, SORTSIZE, SET SASAUTOS, SASINITIALFOLDER.
3. **Does SASAUTOS automatically search subdirectories when a path is specified?** No, each subdirectory must be included in SASAUTOS to retrieve any macros in subdirectories.
4. **When submitting code in batch, is there a method to submit a line of syntax greater than 256 characters?** Yes, by including line breaks to break up the 256-character statement. Another option is to save the long SAS line of code in a separate file and %INCLUDE the code and set the LRECL = option as in this example — FILENAME MYFILE 'PATH TO FILE' LRECL = 32767;
5. **In general, what is a good matching records threshold value to use an index to increase efficiency?** When about 15% of the records match the subset condition, then it makes sense to use indexes to reduce CPU.
6. **What are at least three methods for applying table-look techniques?** Array processing, Hash objects, Formats, and Combining/Merging data.
7. **What are the five main factors that determine a SAS program's efficiency?** CPU, Memory, Input/Output, Data Storage, and Programming Time.
8. **In general, what percent improvement can be expected using simple programming techniques?** About 80% improvement can be made with simple programming techniques.
9. **When applying the compression technique to save space, is it more effective for short and wide data sets or for long and narrow data sets?** Short and wide datasets.

10. **In general, is it more efficient to apply DROP or KEEP as dataset options or SAS statements?** DROP or KEEP as dataset options are more efficient than as SAS statements.
11. **Do SAS views take up any space?** No, they only contain the SAS syntax to create the virtual dataset.
12. **When using PROC SORT, how much disk space is required?** About 2.5 times the dataset space is required to run PROC SORT.
13. **What are at least two methods for creating indexes?** DATA step, PROC DATASETS, and PROC SQL.
14. **Which two DATA step temporary variables are useful in identifying and removing duplicate records?** FIRST. and LAST.

CHAPTER 4: WHAT'S NEW IN SAS VERSION 9.3—ANSWERS

1. **What is the PROC SORT option to make sorts case insensitive?** SORTSEQ = option.
2. **Which function is useful for counting words in a string?** COUNTW()
3. **Which function is useful for locating a word and word number?** FINDW()
4. **Which function is useful for extracting the first character in a string?** FIRST()
5. **What is the difference between CPORT and CIMPORT?** CPORT is used to create transport files from SAS datasets, and CIMPORT is used to create SAS datasets from transport files.
6. **Which CIMPORT options will enable dataset names or member names to be up to 32 bytes in length?** VALIDVARNAME = ANY or VALIDMEMNAME = EXTEND.
7. **Can PROC FCMP be used to create character or numeric functions?** Yes
8. **What is the option that copies to the output dataset the SAS internal date and time when the input dataset was created and the date and time when it was last modified prior to the sort?** DATECOPY
9. **Which function returns the first position of alphanumeric character?** ANYALNUM()
10. **Which new SAS 9.0 function is similar to CAT() plus TRIM() plus LEFT()?** CATS()

Index

SYMBOLS

A

B

C

D

H

I

Q

T

U

V

W

X

Y